# DESIGN OF THERMAL SYSTEMS

## Second Edition

**W. F. Stoecker**

*Professor of Mechanical Engineering*
*University of Illinois at Urbana-Champaign*

**McGraw-Hill Book Company**

New York  St. Louis  San Francisco  Auckland  Bogotá  Hamburg
Johannesburg  London  Madrid  Mexico  Montreal  New Delhi
Panama  Paris  São Paulo  Singapore  Sydney  Tokyo  Toronto

**DESIGN OF THERMAL SYSTEMS**

2 3 4 5 6 7 8 9 0    DODO    8 9 8 7 6 5 4 3 2 1

This book was set in Times Roman by Automated Composition Service, Inc.
The editors were B. J. Clark and Madelaine Eichberg;
the production supervisor was Donna Piligra.
The cover was designed by Scott Chelius.
R. R. Donnelley & Sons Company was printer and binder.

**Library of Congress Cataloging in Publication Data**

Stoecker, Wilbert F
    Design of thermal systems.

    Includes bibliographies and index.
    1. Heat engineering.  2. Systems engineering.
3. Engineering design.  I. Title.
TJ260.S775    1980      621.4′02     79-19425
**ISBN 0-07-061618-3**

# CONTENTS

# PREFACE
# TO THE SECOND EDITION

The field of thermal system design has begun to mature. The origin of the discipline was probably the University of Michigan project sponsored in the mid-1960s by the National Science and Ford Foundations. The motivation at that time might have been considered artificial, because the participants in that program were seeking ways of using digital computers in engineering education. The topics and techniques identified in the project, modeling, simulation, and optimization, proved to be significant.

The first edition of *Design of Thermal Systems* appeared in the early 1970s and concentrated on the applications to thermal systems of modeling, simulation, and optimization. At that time the industrial applications were somewhat rare, essentially limited to large chemical and petroleum facilities.

The emergence of the energy crisis about 1973 provided the impetus for the industrial application of system simulation. System simulation has become an accepted tool for energy analysis of power generating, air conditioning, refrigeration, and other thermal processing plants. Simulation is often used in the design or development stage to evaluate energy requirements of the proposed system or to explore potential savings in first cost. The acceptance of optimization techniques as an industrial tool is moving less rapidly, but many engineers feel that it is only a matter of time and increased familiarity with the power of sophisticated optimization techniques before acceptance becomes widespread.

The reason for preparing a second edition of *Design of Thermal Systems* is that the field has advanced during the years since the appearance of the first edition. Some of the techniques of simulation and optimization have become more stabilized. The author also believes that some of the topics can now be explained more clearly and that additional examples and problems can vitalize the use of these topics. Since the comprehensive designs at the end of the text have been attractive to many instructors, their number has been increased.

*W. F. Stoecker*

# PREFACE
# TO THE FIRST EDITION

The title, *Design of Thermal Systems*, reflects the three concepts embodied in this book: *design*, *thermal*, and *systems*.

## DESIGN

A frequent product of the engineer's efforts is a drawing, a set of calculations, or a report that is an abstraction and description of hardware. Within engineering education, the cookbook approach to design, often practiced during the 1940s, discredited the design effort so that many engineering schools dropped design courses from their curricula in the 1950s. But now design has returned. This reemergence is not a relapse to the earlier procedures; design is reappearing as a creative and highly technical activity.

## THERMAL

Within many mechanical engineering curricula the term *design* is limited to *machine design*. In order to compensate for this frequent lack of recognition of thermal design, some special emphasis on this subject for the next few years is warranted. The designation *thermal* implies calculations and activities based on principles of thermodynamics, heat transfer, and fluid mechanics.

The hardware associated with thermal systems includes fans, pumps, compressors, engines, expanders, turbines, heat and mass exchangers, and reactors, all interconnected with some form of conduits. Generally, the working substances are fluids. These types of systems appear in such industries as power generation, electric and gas utilities, refrigeration, air conditioning and heating, and in the food, chemical, and process industries.

## SYSTEMS

Engineering education is predominantly *process oriented*, while engineering practice is predominantly *system oriented*. Most courses of study in engineering provide the student with an effective exposure to such processes as the flow of a compressible fluid through a nozzle and the behavior of hydrodynamic and thermal boundary layers at solid surfaces. The practicing engineer, however, is likely to be confronted with a task such as designing an economic system that receives natural gas from a pipeline and stores it underground for later usage. There is a big gap between knowledge of individual processes and the integration of these processes in an engineering enterprise.

Closing the gap should not be accomplished by diminishing the emphasis on processes. A faulty knowledge of fundamentals may result in subsequent failure of the system. But within a university environment, it is beneficial for future engineers to begin thinking in terms of systems. Another reason for more emphasis on systems in the university environment, in addition to influencing the thought patterns of students, is that there are some techniques—such as simulation and optimization—which only recently have been applied to thermal systems. These are useful tools and the graduate should have some facility with them.

While the availability of procedures of simulation and optimization is not a new situation, the practical application of these procedures has only recently become widespread because of the availability of the digital computer. Heretofore, the limitation of time did not permit hand calculations, for example, of an optimization of a function that was dependent upon dozens or hundreds of independent variables. This meant that, in designing systems consisting of dozens or hundreds of components, the goal of achieving a *workable* system was a significant accomplishment and the objective of designing an *optimum* system was usually abandoned. The possibility of optimization represents one of the few facets of design.

## OUTLINE OF THIS BOOK

The goal of this book is the design of optimum thermal systems. Chapters 6 through 11 cover topics and specific procedures in optimization. After Chap. 6 explains the typical statement of the optimization problem and illustrates how this statement derives from the physical situation, the chapters that follow explore optimization procedures such as calculus methods, search methods, geometric programming, dynamic programming, and linear programming. All these methods have applicability to many other types of problems besides thermal ones and, in this sense, are general. On the other hand, the applications are chosen from the thermal field to emphasize the opportunity for optimization in this class of problems.

If the engineer immediately sets out to try to optimize a moderately complex thermal system, he is soon struck by the need for predicting the performance of that system, given certain input conditions and performance characteristics of components. This is the process of *system simulation.* System simulation not only may be a step in the optimization process but may have a usefulness in its own right. A system may be designed on the basis of some maximum load condition but may operate 95 percent of the time at less-than-maximum load. System simulation permits an examination of the operating conditions that may pinpoint possible operating and control problems at non-design conditions.

Since system simulation and optimization on any but the simplest problems are complex operations, the execution of the problem must be performed on a computer. When using a computer, the equation form of representation of the performance of components and expression of properties of substances is much more convenient than tabular or graphical representations. Chapter 4 on mathematical modeling presents some techniques for equation development for the case where there is and also where there is not some insight into the relationships based in thermal laws.

Chapter 3, on economics, is appropriate because engineering design and economics are inseparable, and because a frequent criterion for optimization is the economic one. Chapter 2, on workable systems, attempts to convey one simple but important distinction—the difference between the design process that results in a workable system in contrast to an optimum system. The first chapter on engineering design emphasizes the importance of design in an engineering undertaking.

The appendix includes some problem statements of several comprehensive projects which may run as part-time assignments during an entire term. These term projects are industrially oriented but require application of some of the topics explained in the text.

The audience for which this book was written includes senior or first-year graduate students in mechanical or chemical engineering, or practicing engineers in the thermal field. The background assumed is a knowledge of thermodynamics, heat transfer, fluid mechanics, and an awareness of the performance characteristics of such thermal equipment as heat exchangers, pumps, and compressors. The now generally accepted facility of engineers to do basic digital computer programming is also a requirement.

## ACKNOWLEDGMENTS

Thermal system design is gradually emerging as an identifiable discipline. Special recognition should be given to the program coordinated by the University of Michigan on Computers in Engineering Design Education, which in 1966 clearly delineated topics and defined directions that have since proved to be productive. Acknowledgment should be given to activities within the chemical engineering field for developments that are closely related, and in some cases identical, to those in the thermal stem of mechanical engineering.

Many faculty members during the past five years have arrived, often independently, at the same conclusion as the author: the time is opportune for developments in thermal design. Many of these faculty members have shared some of their experiences in the thermal design section of *Mechanical Engineering News* and have, thus, directly and indirectly contributed to ideas expressed in this book.

This manuscript is the third edition of text material used in the Design of Thermal Systems course at the University of Illinois at Urbana-Champaign. I thank the students who have worked with me in this course for their suggestions for improvement of the manuscript. The second edition was an attractively printed booklet prepared by my Department Publication Office, George Morris, Director; June Kempka and Dianne Merridith, typists; and Don Anderson, Bruce Breckenfeld, and Paul Stoecker, draftsmen. Special thanks are due to the Engineering Department of Amoco Chemicals Corporation, Chicago, for their interest in engineering education and for their concrete evidence of this interest shown by printing the second edition.

Competent colleagues are invaluable as sounding boards for ideas and as contributors of ideas of their own. Professor L. E. Doyle offered suggestions on the economics chapter and Prof. C. O. Pedersen, a coworker in the development of the thermal systems program at the University of Illinois at Urbana-Champaign, provided advice at many stages. Mr. Donald R. Witt and a class of architectural engineering students at Pennsylvania State University class-tested the manuscript and provided valuable suggestions from the point of view of a user of the book. Beneficial comments and criticisms also came from the Newark College of Engineering, where Prof. Eugene Stamper and a group of students tested the manuscript in one of their classes. Professor Jack P. Holman of Southern Methodist University, consulting editor of McGraw-Hill Book Company, supplied perceptive comments both in terms of pedagogy as well as in the technical features of thermal systems.

The illustrations in this book were prepared by George Morris of Champaign, Illinois.

By being the people that they are, my wife Pat and children Paul, Janet, and Anita have made the work on this book, as well as anything else that I do, seem worthwhile.

*W. F. Stoecker*

# ONE

## ENGINEERING DESIGN

## 1-1 INTRODUCTION

Typical professional activities of engineers include sales, construction, research, development, and design. Design will be our special concern in this book. The immediate product of the design process is a report, a set of calculations, and/or a drawing that are abstractions of hardware. The subject of the design may be a process, an element or component of a larger assembly, or an entire system.

Our emphasis will be on *system* design, where a system is defined as a collection of components with interrelated performance. Even this definition often needs interpretation, because a large system sometimes includes subsystems. Furthermore, we shall progressively focus on *thermal* systems, where fluids and energy in the form of heat and work are conveyed and converted. Before adjusting this focus, however, this chapter will examine the larger picture into which the technical engineering activity blends. We shall call this larger operation an engineering *undertaking*, implying that engineering plays a decisive role but also dovetails with other considerations. Engineering undertakings include a wide variety of commercial and industrial enterprises as well as municipally, state-, and federally sponsored projects.

## 1-2 DECISIONS IN AN ENGINEERING UNDERTAKING

In recent years an appreciable amount of attention has been devoted to the *methodology* or the *morphology* of engineering undertakings. Studies on these topics have analyzed the steps and procedures used in reaching decisions. One

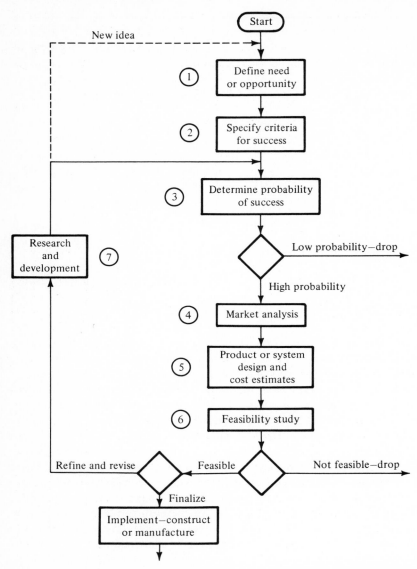

**Figure 1-1** Possible flow diagram in evaluating and planning an engineering undertaking.

contribution of these studies has been to stimulate engineers to reflect on the thinking processes of themselves and others on the project team. Certainly the process and sequence of steps followed in each undertaking is different, and no one sequence, including the one described in this chapter, is universally applicable. Since the starting point, the goal, and the side conditions differ from one undertaking to the next, the procedures must vary.

The advantage of analyzing the decision process, especially in complex undertakings, is that it leads to a more logical coordination of the many individual efforts constituting the entire venture. The flow diagram in Fig. 1-1 shows typical steps followed in the conception, evaluation, and execution of the plan. The rectangular boxes, which indicate actions, may represent considerable effort and expenditures on large projects. The diamond boxes represent decisions, e.g., whether to continue the project or to drop it.

The technical engineering occurs mostly in activities 5 and 7, product or system design and research and development. Little will be said in this chapter about product or system design because it will be studied in the chapters to follow. The flow diagram shows only how this design procedure fits into the larger pattern of the undertaking. The individual nondesign activities will be discussed next.

## 1-3  NEED OR OPPORTUNITY (STEP 1)

Step 1 in the flow diagram of Fig. 1-1 is to define the need or opportunity. It may seem easy to state the need or opportunity, but it is not always a simple task. For example, the officials of a city may suppose that their need is to enlarge the reservoir so that it can store a larger quantity of water for municipal purposes. The officials may not have specified the actual need but have leaped to one possible solution. Perhaps the need would better have been stated as a low water reserve during certain times of the year. Enlargement of the reservoir might be one possible solution, but other solutions might be to restrict the consumption of water and to seek other sources such as wells. Sometimes possible solutions are precluded by not stating the need properly at the beginning.

The word "opportunity" has positive connotations, whereas "need" suggests a defensive action. Sometimes the two cannot be distinguished. For example, an industrial firm may recognize a new product as an opportunity, but if the company does not then expand its line of products, business is likely to decline. Thus the introduction of a new product is also a need.

In commercial enterprises, typical needs or opportunities lie in the renovation or expansion of facilities to manufacture or distribute a current product. Opportunity also arises when the sale of a product not manufactured by the firm is rising and the market potential seems favorable. Still a third form in which an opportunity arises is through research and development within the organization. A new product may be developed intentionally or accidentally. Sometimes a new use of an existing product can be found by making a slight modification of it. An organization may know how to manufacture a gummy, sticky substance and assign to the research and development department the task of finding some use for it.

Of interest to us at the moment is the need or opportunity that requires engineering design at a subsequent stage.

## 1-4 CRITERIA OF SUCCESS (STEP 2)

In commercial enterprises the usual criterion of success is showing a profit, i.e., providing a certain rate of return on the investment. In public projects the criterion of success is the degree to which the need is satisfied in relation to the cost, monetary or otherwise.

In a profit-and-loss economy, the expected earning power of a proposed commercial project is a dominating influence on the decision to proceed with the project. Strict monetary concerns are always tempered, however, by human, social, and political considerations to a greater or lesser degree. In other words, a price tag is placed on the nonmonetary factors. A factory may be located at a more remote site at a penalty in the form of transportation costs so that its atmospheric pollution or noise affects fewer people. As an alternative, the plant may spend a lot on superior pollution control in order to be a good neighbor to the surrounding community.

Sometimes a firm will design and manufacture a product that offers little opportunity for profit simply to round out a line of products. The availability of this product, product $A$, permits the sales force to say to a prospective customer, "Yes, we can sell you product $A$, but we recommend product $B$," which is a more profitable item in the company's line and may actually be superior to product $A$.

Often a decision, particularly in an emergency, appears outside the realm of economics. If a boiler providing steam for heating a rental office building fails, the decision whether to repair or replace the boiler may seem to be outside the realm of economics. The question can still be considered an economic one, however, the penalty for not executing the project being an overpowering loss.

## 1-5 PROBABILITY OF SUCCESS (STEP 3)

Plans and designs are always directed toward the future, for which only probability, not certainty, is applicable. There is no absolute assurance that the plant will meet the success criteria discussed in Sec. 1-4, only a likelihood or probability that it will do so.

The mention of *probability* suggests the normal distribution curve (Fig. 1-2), an excellent starting point for expressing uncertainty in the decision-making process. The significance of the distribution curve lies particularly in the evaluation of the area beneath the curve. The area under the curve between $x_1$ and $x_2$, for example, represents the probability $P$ of the event's occurring between values $x_1$ and $x_2$. Thus,

$$P = \int_{x_1}^{x_2} y \, dx$$

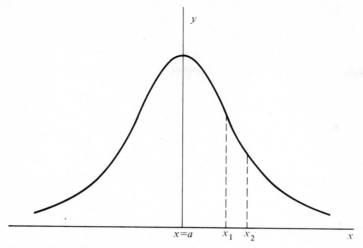

**Figure 1-2** Probability distribution curve.

Since the probability of the event's occurring somewhere in the range of $x$ is unity, the integration over the entire range of $x$ is equal to 1.0:

$$\int_{-\infty}^{\infty} y\,dx = 1$$

The equation for the probability distribution curve is

$$y = \frac{h}{\sqrt{\pi}}\, e^{-h^2\,(x-a)^2} \tag{1-1}$$

The maximum value of the ordinate is $h/\sqrt{\pi}$, which occurs when $x = a$. This fact suggests that increasing the value of $h$ alters the shape of the distribution curve, as shown in Fig. 1-3. If $h_1$ is greater than $h_2$, the peak of the $h_1$ curve rises higher than that of the $h_2$ curve.

To extend the probability idea to decision making in an engineering undertaking, suppose that a new product or facility is proposed and that the criterion for success is a 10 percent rate of return on the investment for a 5-year life of the plant. After a preliminary design, the probability distribution curve is shown as indicated in Fig. 1-4. Since rough figures were used throughout the evaluation, the distribution curve is flat, indicating no great confidence in an expected percent of return of investment of, say, 18 percent. The expected rate of return is attractive enough, however, to proceed to a complete design, including cost estimates. If the most probable return on investment after this complete design were 16 percent, for example, the confidence in this figure would be greater than the confidence in the 18 percent figure after the preliminary design because costs

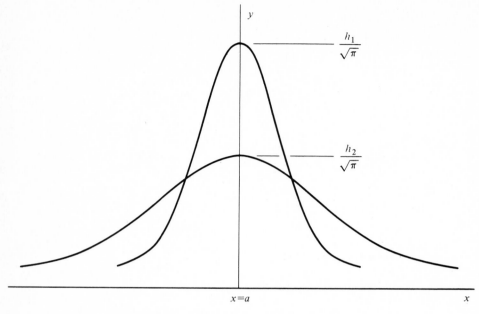

**Figure 1-3** Several different shapes of the probability distribution curve.

have now been analyzed more carefully and marketing studies have been conducted more thoroughly.

The probability distribution curves at two other stages, after construction and after 1 year of operation, show progressively greater degrees of confidence in the rate of return after a 5-year life. After 5 years, the rate of return is known exactly, and the probability distribution curve degenerates into a curve that is infinitesimally thin and infinitely high.

The recognition that prediction of future behavior is not deterministic, so that only one set of events or conditions will prevail, has spawned a new probabilistic approach to design (see the additional readings at the end of the chapter). One of the activities of this new study is that of quantifying the curves shown in Fig. 1-4. It is valuable for the decision maker to know not only the most likely value of the return on investment but also whether there is a high or low probability of achieving this most likely value.

## 1-6 MARKET ANALYSIS (STEP 4)

If the undertaking is one in which a product or service must eventually be sold or leased to customers, there must be some indication of favorable reaction by the potential consumer. An ideal form of the information provided by a market analysis would be a set of curves like those in Fig. 1-5. With an increase in price,

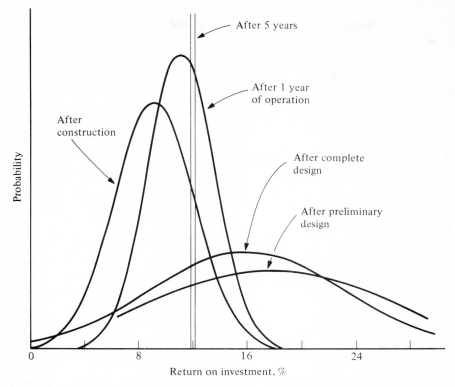

**Figure 1-4** Distribution curves at various stages of decision making.

the potential volume of sales decreases until such a high price is reached that no sales can be made. The sales-volume to price relationship affects the size of the plant or process because the unit price is often lower in a large plant. For this reason, the marketing and plant capabilities must be evaluated in conjunction with each other.

Because the sales and advertising effort influences the volume of sales for a given price, a family of curves is expected. Since a cost is associated with the sales and advertising effort, and since a continuous increase of this effort results in diminishing improvement in sales, there exists an optimum level of sales and advertising effort. A marketing plan should emerge simultaneously with the technical plans for the undertaking.

## 1-7 FEASIBILITY (STEP 6)

The feasibility study, step 6, and the subsequent feasibility decision refer to whether the project is even possible. A project may be feasible, or possible, but not economical. Infeasibility may result from unavailability of investment

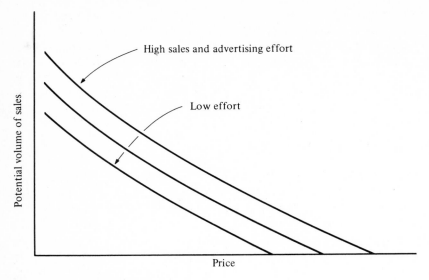

**Figure 1-5** End result of a market analysis.

capital, land, labor, or favorable zoning regulations. Safety codes or other regu-
latory laws may prohibit the enterprise. If an undertaking is shown to be infeasible, either alternatives must be found or the project must be dropped.

## 1-8 RESEARCH AND DEVELOPMENT (STEP 7)

If the product or process is one new to the organization, the results from research and development (R&D) may be an important input to the decision
process. Research efforts may provide the origin or improvement of the basic
idea, and development work may supply working models or a pilot plant, depending upon the nature of the undertaking.

Placing R&D in a late stage of decision making, as was done in Fig. 1-1,
suggests that an idea originates somewhere else in the organization or in the
field and eventually is placed at the doorstep of R&D for transformation into a
workable idea. The possibility of the idea's originating in the research group
should also be exploited and is indicated by the dashed line in Fig. 1-1. Research
people often learn of new ideas in other fields which might be applied to their
own activity.

## 1-9 ITERATIONS

The loop in Fig. 1-1 emphasizes that the decision-making process involves many
iterations. Each pass through the loop improves the amount and the quality of

information and data. Eventually a point is reached where final decisions are made regarding the design, production, and marketing of the product. The substance that circulates through this flow diagram is information, which may be in the form of reports and conversations and may be both verbal and pictorial. The iterations are accomplished by communication between people, and this communication is interspersed by go-or-no-go decisions.

## 1-10 OPTIMIZATION OF OPERATION

The flow diagram of Fig. 1-1 terminates with the construction or beginning of manufacture of a product or service. Actually another stage takes over at this point, which seeks to optimize the operation of a given facility. The facility was designed on the basis of certain design parameters which almost inevitably change by the time the facility is in operation. The next challenge, then, is to operate the facility in the best possible manner in the light of such factors as actual costs and prices. A painful activity occurs when the project is not profitable and the objective becomes that of minimizing the loss.

## 1-11 TECHNICAL DESIGN (STEP 5)

Step 5 in Fig. 1-1, the product or system design, has not been discussed. The reason for this omission is that the system design is the subject of this book from this point on. This step is where the largest portion of engineering time is spent. System design as an activity lies somewhere between the study and analysis of individual processes or components and the larger decisions, which are heavily economic. Usually one person coordinates the planning of the undertaking. This manager normally emerges with a background gained from experience in one of the subactivities. The manager's experience might be in finance, engineering, or marketing, for example. Whatever the original discipline, the manager must become conversant with all the fields that play a role in the decision-making process.

The word "design" encompasses a wide range of activities. Design may be applied to the act of selecting a single member or part, e.g., the size of a tube in a heat exchanger; to a larger component, e.g., the entire shell-and-tube heat exchanger; or to the design of the system in which the heat exchanger is only one component. Design activities can be directed toward mechanical devices which incorporate linkages, gears, and other moving solid members, electrical or electronic systems, thermal systems, and a multitude of others. Our concentration will be on thermal systems such as those in power generation, heating and refrigeration plants, the food-processing industry, and in the chemical and process industries.

## 1-12 SUMMARY

The flow diagram and description of the decision processes discussed in this chapter are highly simplified and are not sacred. Since almost every undertaking is different, there are almost infinite variations in starting points, goals, and intervening circumstances. The purpose of the study is to emphasize the advantage of systematic planning. Certain functions are common in the evaluation and planning of undertakings, particularly the iterations and the decisions that occur at various stages.

## ADDITIONAL READINGS

**Introductory books on engineering design**

Alger, J. R. M., and C. V. Hays: *Creative Synthesis in Design*, Prentice-Hall, Englewood Cliffs, N.J., 1964.
Asimow, M.: *Introduction to Design*, Prentice-Hall, Englewood Cliffs, N.J., 1962.
Beakley, G. C., and H. W. Leach: *Engineering, An Introduction to a Creative Profession*, Macmillan, New York, 1967.
Buhl, H. R.: *Creative Engineering Design*, Iowa State University Press, Ames, 1960.
Dixon, J. R.: *Design Engineering: Inventiveness, Analysis, and Decision Making*, McGraw-Hill, New York, 1966.
Harrisberger, L.: *Engineersmanship, A Philosophy of Design*, Brooks/Cole, Belmont, Calif., 1966.
Krick, E. V.: *An Introduction to Engineering and Engineering Design*, Wiley, New York, 1965.
Middendorf, W. H.: *Engineering Design*, Allyn and Bacon, Boston, 1968.
Mischke, C. R.: *An Introduction to Computer-Aided Design*, Prentice-Hall, Englewood Cliffs, N.J., 1968.
Morris, G. E.: *Engineering, A Decision-Making Process*, Houghton Mifflin Company, Boston, 1977.
Woodson, T. T.: *Introduction to Engineering Design*, McGraw-Hill, New York, 1966.

**Probabilistic approaches to design**

Ang, A. H-S., and W. H. Tang: *Probability Concepts in Engineering Planning and Design*, Wiley, New York, 1975.
Haugen, E. B.: *Probabilistic Approaches to Design*, Wiley, New York, 1968.
Rudd, D. F., and C. C. Watson: *Strategy of Process Engineering*, Wiley, New York, 1968.

# TWO

## DESIGNING A WORKABLE SYSTEM

## 2-1 WORKABLE AND OPTIMUM SYSTEMS

The simple but important point of this chapter is the distinction between designing a workable system and an optimum system. This chapter also continues the progression from the broad concerns of an undertaking, as described in Chap. 1, to a concentration on engineering systems and, even more specifically, on thermal systems.

It is so often said that "there are many possible answers to a design problem" that the idea is sometimes conveyed that all solutions are equally desirable. Actually only one solution is the optimum, where the optimum is based on some defined criterion, e.g., cost, size, or weight. The distinction then will be made between a workable and an optimum system. It should not be suggested that a workable system is being scorned. Obviously, a workable system is infinitely preferable to a nonworkable system. Furthermore, extensive effort in progressing from a workable toward an optimum system may not be justified because of limitations in calendar time, cost of engineering time, or even the reliability of the fundamental data on which the design is based. One point to be explored in this chapter is how superior solutions may be ruled out in the design process by prematurely eliminating some system concepts. Superior solutions may also be precluded by fixing interconnecting parameters between components and selecting the components based on these parameters instead of letting the parameters float until the optimum total system emerges.

## 2-2 A WORKABLE SYSTEM

A workable system is one that

1. Meets the requirements of the purposes of the system, e.g., providing the required amount of power, heating, cooling, or fluid flow, or surrounding a space with a specified environment so that people will be comfortable or a chemical process will proceed or not proceed
2. Will have satisfactory life and maintenance costs
3. Abides by all constraints, such as size, weight, temperatures, pressure, material properties, noise, pollution, etc.

In summary, a workable system performs the assigned task within the imposed constraints.

## 2-3 STEPS IN ARRIVING AT A WORKABLE SYSTEM

The two major steps in achieving a workable system are (1) to select the concept to be used and (2) to fix whatever parameters are necessary to select the components of the system. These parameters must be chosen so that the design requirements and constraints are satisfied.

## 2-4 CREATIVITY IN CONCEPT SELECTION

Engineering, especially engineering design, is a potentially creative activity. In practice creativity may not be exercised because of lack of time for adequate exploration, discouragement by supervision or environment, or the laziness and timidity of the engineer. It is particularly in selecting the concept that creativity can be exercised. Too often only one concept is ever considered, the concept that was used on the last similar job. As a standard practice, engineers should discipline themselves to review all the alternative concepts in some manner appropriate to the scope of the project. Old ideas that were once discarded as impractical or uneconomical should be constantly reviewed. Costs change; new devices or materials on the market may make an approach successful today that was not attractive 10 years ago.

## 2-5 WORKABLE VS. OPTIMUM SYSTEM

The distinction between the approaches used in arriving at a workable system and an optimum system can be illustrated by a simple example. Suppose that the pump and piping are to be selected to convey 3 kg/s from one location to another 250 m away from the original position and 8 m higher. If the design is approached with the limited objective of achieving a workable system, the following procedure might be followed:

1. The elevation of 8 m imposes a pressure difference of

$$(8 \text{ m}) (1000 \text{ kg/m}^3) (9.807 \text{ m/s}^2) = 78.5 \text{ kPa}$$

   Arbitrarily choose an additional 100 kPa to compensate for friction in the 250 m of pipe.
2. According to the foregoing decision, select a pump which delivers 3 kg/s against a pressure difference of 178.5 kPa. Finally, select a pipe size from a handbook such that the pressure drop in 250 m of length is 100 kPa or less. A pipe size of 50 mm (2 in) satisfies the requirement.

Approaching the same problem with the objective of achieving an optimum system presupposes agreement on a criterion to optimize. A frequently chosen criterion is cost (sometimes first cost only in speculative projects, and sometimes the lifetime cost, consisting of first plus lifetime pumping and maintenance costs).

In designing the optimum pump and piping system for minimum lifetime cost, the pressure rise to be developed by the pump is not fixed immediately but left free to float. If the three major contributors to cost are (1) the first cost of the pump, (2) the first cost of the pipe, and (3) the lifetime pumping costs, these costs will vary as a function of pump pressure, as shown in Fig. 2-1. As the pump-pressure rise increases, the cost of the pump probably increases for the required flow rate of 3 kg/s because of the need for higher speed and/or larger impeller diameter. With the increase in pressure rise, the power required by the pump increases and is reflected in a higher lifetime pumping cost. The first cost of the pipe, the third contributor to the total cost, becomes enormously high as the pressure available to overcome friction in the pipe reduces to zero. The available pressure for the pipe is the pump-pressure rise minus 78.5 kPa needed for the difference in elevation. An appropriate optimization technique can be used to determine the optimal pump-pressure rise, which in Fig. 2-1 is approximately 125 kPa. Finally the pump can be selected to develop 125 kPa pressure rise, and a pipe size can be chosen such that the pressure drop due to friction is 46.5 kPa or less.

The tone of the preceding discussion indicates a strong preference toward designing optimum systems. To temper this bias, several additional considerations should be mentioned. If the job is a small one, the cost of the increased engineering time required for optimization may devour the savings, if any. Not only the engineer's time but pressure of calendar time may not permit the design to proceed beyond a workable design.

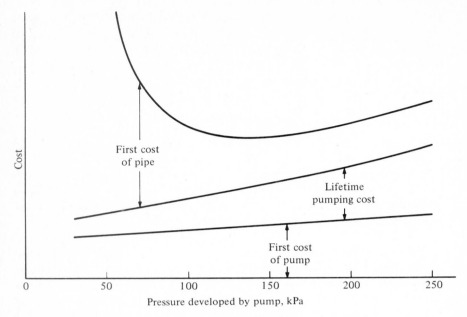

**Figure 2-1** Contributions to costs by pump and piping systems.

## 2-6  DESIGN OF A FOOD-FREEZING PLANT

Large-scale engineering projects are extremely complex, and decisions are often intricately interrelated; not only do they influence each other in the purely technical area but also cross over into the technoeconomic, social, and human fields. To illustrate a few of the decisions involved in a realistic commercial undertaking and to provide a further example of the contrast between a workable system and an optimum system, consider the following project.

A food company can buy sweet corn and peas from farmers during the season and sell the vegetables as frozen food throughout the year in a city 300 km away. What are the decisions and procedures involved in designing the plant to process and freeze the crops?

The statement of the task actually starts at an advanced stage in the decision process, because it is already assumed that a plant will be constructed. This decision cannot realistically be made until some cost data are available to evaluate the attractiveness of the project. Let us assume, therefore, that an arbitrarily selected solution has been priced out and found to be potentially profitable. We are likely, then, to arrive at a solution that is an improvement over the arbitrary selection.

Some major decisions that must be made are (1) the location, (2) size, and (3) type of freezing plant. The plant could be located near the producing area, in the market city, or somewhere between. The size will be strongly influenced by the market expectation. The third decision, the type of freezing plant, embraces the engineering design. These three major decisions are interrelated. For example, the location and size of plant might reasonably influence the type of system selected. The selection of the type of freezing plant includes choosing the concept on which the freezing-plant design will be based. After the concept has been decided, the internal design of the plant can proceed.

An outline of the sequence of tasks and decisions by which a workable design could be arrived at is as follows:

1. Decide to locate the plant in the market city adjacent to a refrigerated warehouse operated by the company.
2. Select the freezing capacity of the plant on the basis of the current availability of the crop, the potential sale in the city, and available financing.
3. Decide upon the concept to be used in the freezing plant, e.g., the one shown in Fig. 2-2. In this system the food particles are frozen in a fluidized bed, in which low-temperature air blows up through a conveyor chain, suspending the product being frozen. This air returns from the fluidized-bed conveyor to a heat exchanger that is the evaporator of a refrigerating unit. The refrigerating unit uses a reciprocating compressor and water-cooled condenser. A cooling tower, in turn, cools the condenser water, rejecting heat to the atmosphere.
4. The design can be quantified by establishing certain values. Since the throughput of the plant has already been determined, the freezing capacity in kilograms per second can be computed by deciding upon the number of shifts to

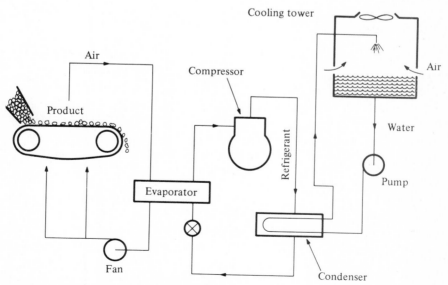

**Figure 2-2** Schematic flow diagram of freezing plant.

**Table 2-1**

|                                   | Temperature, °C |
| --------------------------------- | --------------- |
| Air, chilled supply               | −30             |
| Return                            | −23             |
| Refrigerant, evaporation          | −38             |
| Condensation                      | 45              |
| Condenser cooling water, inlet    | 30              |
| Outlet                            | 35              |

be operated. Assume that one shift is selected, so that now the refrigeration load can be calculated at, say, 220 kW. To proceed with the design, the parameters shown in Table 2-1 can be pinned down.

5. After these values have been fixed, the individual components can be selected. The flow of chilled air can be calculated to remove 220 kW with a temperature rise of 7°C. The conveyor length and speed must now be chosen to achieve the required rate of heat transfer. The air-cooling evaporator can be selected from a catalog because the airflow rate, air temperatures, and refrigerant evaporating temperature fix the choice. The compressor must provide 220 kW of refrigeration with an evaporating temperature of −38°C and a condensing temperature of 45°C, which is adequate information for selecting the compressor or perhaps a two-stage compression system. The heat-rejection rate at the condenser exceeds the 220-kW refrigeration capacity by the amount of work added in the compressor and may be in the neighborhood of 300 kW. The condenser and cooling tower can be sized on the basis of the rate of heat flow and the water temperatures of 30 and 35°C. Thus, a workable system can be designed.

Unlike the above procedure, an attempt to achieve an optimum system returns to the point where the first decisions are made. Such decisions as the location, size, and freezing concept should be considered in connection with each other instead of independently. The choice of fluidized-bed freezing with a conventional refrigeration plant is only one of the commercially available concepts, to say nothing of the possibility (admittedly remote) of devising an entirely new concept. Other concepts are a freezing tunnel, where the air blows over the top of the product; packaging the product first and immersing the package in cold brine until frozen; or freezing the product with liquid nitrogen purchased in liquid form in bulk. An example of the interconnection of decisions is that the location of the plant that is best for one concept may not be best for another concept. A compression refrigeration plant may be best located in the city as an extension of existing freezing facilities, and it may be unwise to locate it close to the producing area because of lack of trained operators. The liquid-nitrogen freezing plant, on the other hand, is simple in operation and could be located close to the field; furthermore, it could be shut down for the idle off-season more conveniently than the compression plant. If the possibility of two or even three shifts were considered, the processing rate of the plant could be reduced by a factor of 2 or 3, respectively, for the same daily throughput.

Within the internal design of the compression refrigeration plant, the procedure was to select reasonable temperatures and then design each component around those temperatures and resulting flow rates. When one approaches the design with the objective of optimization, all those interconnecting parameters are left free to float and one finds the combination of values of these parameters which results in the optimum (probably the economic optimum).

## 2-7 PRELIMINARIES TO THE STUDY OF OPTIMIZATION

Any attempt to apply optimization theory to thermal systems at this stage is destined for frustration. There are a variety of optimization techniques available, some of which are studied in Chaps. 8 to 12. The first attempts to optimize a thermal system with a dozen components, however, will be detoured by the need to predict the performance of the system with given input conditions. This assignment, called *system simulation*, must be studied first and will be considered in Chap. 6. For complex systems, system simulation must be performed with a computer. For this purpose the performance characteristics making up the system could possibly be stored as tables, but a far more efficient and useful form is equation-type formulation. Translating catalog tables into equations, called *component simulation* or *mathematical modeling*, is a routine preliminary step to system simulation and will be treated in Chaps. 4 and 5.

Finally, since optimization presupposes a criterion, which in engineering practice is often an economic one, a review of investment economics in Chap. 3 will be appropriate.

The sequence to be followed in the ensuing studies, then, will be (1) economics, (2) mathematical modeling, (3) system simulation, and (4) optimization.

## PROBLEMS

**2-1** Location $S$ in Fig. 2-3 is an adequate source of water, and locations $A$, $B$, and $C$ are points at which water must be provided at the following rates of flow:

| Location | $A$ | $B$ | $C$ |
|----------|-----|-----|-----|
| L/s      | 2.5 | 3.5 | 1.5 |

Points $S$, $A$, $B$, and $C$ are all at the same elevation. The demands for water at $A$ and $C$ occur intermittently and only during the working day, and they may coincide. The demand for water at $B$ occurs only during nonworking hours and is also intermittent. Ground-level access exists in a 3-m border surrounding the building. Access is not permitted over, through, or under the building.

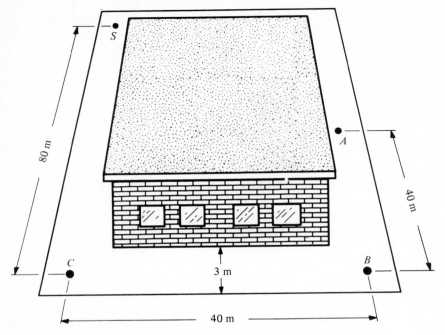

**Figure 2-3** Supply and consumption points in water-distribution system.

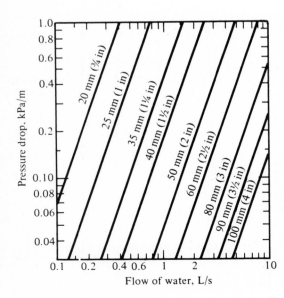

**Figure 2-4** Pressure drop in pipe.

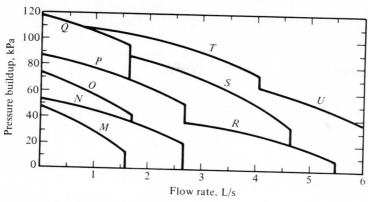

**Figure 2-5** Pump performance curves.

(*a*) Describe all the concepts of workable methods you can devise to fulfill the assignment.

(*b*) The influence of such factors as the expected life of the system has resulted in the decision to use a system in which a pump delivers water into an elevated storage tank, which supplies the piping system. A water-level switch starts and stops the pump. Design the system; this includes sketching the pipe network chosen, listing all the pipe sizes, selecting the pump, and specifying the elevation of the storage tank. Use pressure-drop data from Fig. 2-4 and pump performance from Fig. 2-5. (Neglect the pressure drop in the pipe fittings and pressure conversions due to kinetic energy.) Fill out Table 2-2.

(*c*) Review the design and list the decisions that preclude possible optimization later in the design.

**Table 2-2  Design data for Prob. 2-1**

| Pipe section | Pipe size, mm | Design flow, L/s | | $\Delta p$, kPa | |
| --- | --- | --- | --- | --- | --- |
| | | Day | Night | Day | Night |
| (*S* to *A*, for example) | | | | | |
| | | | | | |
| | | | | | |
| | | | | | |

Pump selected _____    Elevation of water in storage tank _____

**2-2** A heating and ventilating system for a public indoor swimming pool is to be designed.

## Specifications

Pool water temperature, 26°C
Indoor air dry-bulb temperature, 27°C
Outdoor design temperature, −12°C
Outdoor design humidity ratio, 0.00105 kg/kg
For odor control minimum rate of outdoor air for ventilation, 5.8 kg/s
For comfort, temperature of air introduced, 38 to 65°C

## Construction features

Pool dimensions, 15 × 45 m
Essentially the structure has masonry walls and a glass roof:
  Glass area, 1850 m$^2$
  Wall area, 670 m$^2$
  $U$ value of wall, 1.2 W/(m$^2 \cdot$ K)

## Supplementary information

Rate of evaporation of water from pool, g/s = (0.04) (area, m$^2$) ($p_w - p_a$)

where $p_w$ = water-vapor pressure at pool temperature, kPa
  $p_a$ = partial pressure of water vapor in surrounding air, kPa

A choice of the type of window must be made (single, double, or triple pane). They have the following heat-transfer coefficients:

| | Heat-transfer coefficent, W/(m$^2 \cdot$ K) | | |
| --- | --- | --- | --- |
| | Single | Double | Triple |
| Outside air film | 34 | 34 | 34 |
| Glass (between external and internal surfaces) | 118 | 8 | 3.4 |
| Inside film | 11.4 | 11.4 | 11.4 |

The basic purpose of the system is as follows:

1. To maintain the indoor temperature when design conditions prevail outdoors
2. To abide by the constraints
3. To prevent condensation of water vapor on the inside of the glass

(a) Describe at least two different concepts for accomplishing the objectives of the design. Use schematic diagrams if useful.

(b) Assume that the concept chosen is one where outdoor air is drawn in, heated, and introduced to the space and an equal quantity of moist room air is exhausted. Should additional heating be required, it is provided either by heating recirculated air or by using convectors around the perimeter of the building. Perform the design calculations in order to specify the following:

Flow rate of ventilation air _____ kg/s
Temperature of air entering space _____°C

Type of glass selected _____
Temperature of inside surface of glass _____ °C
Condition of indoor air: dew point _____ °C
Rate of water evaporated from pool _____ kg/s
Heat loss by conduction through walls and glass _____ kW
Heat supplied to raise temperature of ventilation air from room temperature to supply temperature _____ kW
Heat to recirculated air or perimeter convectors _____ kW

(c) Review the design and list the decisions that precluded possible optimization later in the design.

**2-3** You have just purchased a remote uninhabited island where you plan to give parties lasting till late at night. You need to install an electric power system that will provide at least 8 kW of lighting.

(a) List and describe, in several sentences, three methods of generating power in this remote location, assuming that equipment and supplies can be transported to the island.

(b) Assume that the decision has been made to produce the electric power by means of an engine-driven generator. Specifically, an engine will be direct-connected to a generator that delivers power to bulbs. Available choices are as follows:

*Bulbs.* 100-W bulbs at 115 V. The bulbs may be connected in parallel, in series, or in combination, and the current flowing through a bulb must be above 70 percent of the rated current to obtain satisfactory lighting efficiency but below 110 percent of the rated current to achieve long life.

*Engines.* A choice can be made between two engines whose power deliveries at wide-open throttle are as follows:

| Engine | Power delivery, kW, at given speed, r/s | | | | |
|---|---|---|---|---|---|
| | 10 | 20 | 30 | 40 | 50 |
| 1 | 2.2 | 5.5 | 9.0 | 11.2 | 12.7 |
| 2 | 3.4 | 7.1 | 11.1 | 13.8 | 15.2 |

*Generators.* Single phase, alternating current. No adjustment of output voltage is possible except by varying the speed. The frequency must be greater than 35 Hz in order to prevent flickering of the lights. Neglect the voltage drop and power loss in the transmission lines. A choice can be made between the following two generators:

| Generator | Poles | Efficiency | Maximum allowable current, A | Output voltage, V, at given speed, r/s | | | | |
|---|---|---|---|---|---|---|---|---|
| | | | | 10 | 20 | 30 | 40 | 50 |
| 1 | 2 | 0.72 | 100 | 38 | 71 | 106 | 140 | 174 |
| 2 | 4 | 0.83 | 50 | 59 | 117 | 170 | 224 | 270 |

Make the following design selections and specifications:

Engine number _____
Generator number _____

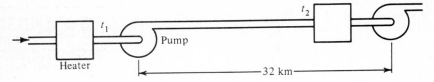

**Figure 2-6** Oil pipeline.

Engine-generator speed _____ r/s
Frequency _____ Hz
Power delivered by engine _____ kW
Power delivered by generator _____ kW
Voltage _____ V
Current _____ A
Number of bulbs and circuiting _____

(c) Review the design and list the decisions that preclude possible optimization later in the design.

**2-4** Crude oil is to be transported overland in Alaska in such a way that the environment is not adversely affected.[1] [†]

(a) Describe two workable methods of transporting this oil.

(b) Of the workable methods, a pipeline is the method chosen for further examination. The inside diameter of the pipe is 600 mm, and the pipe will carry a crude oil flow rate of 44 kg/s. The distance between pumping stations is 32 km. To facilitate pumping, a heater will be installed at each pumping station, as shown in Fig. 2-6. The pipeline is to be buried in permafrost whose temperature at design conditions is −4°C. The permafrost is not to be melted. Insulation may be used on the pipe, and the external surface temperature of the pipe or insulation in contact with the permafrost must be maintained at 0°C or below. Heat-transfer data applicable to the pipe and insulation are shown in Fig. 2-7. The available thicknesses of insulation are 25, 50, 75, and 100 mm. The overall heat-transfer coefficient $U$ between the oil and permafrost in watts per square meter per kelvin based on the inside pipe

[†]Numbered references appear at the end of the chapter.

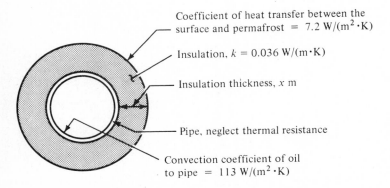

Coefficient of heat transfer between the surface and permafrost = 7.2 W/(m² · K)

Insulation, $k$ = 0.036 W/(m · K)

Insulation thickness, $x$ m

Pipe, neglect thermal resistance

Convection coefficient of oil to pipe = 113 W/(m² · K)

**Figure 2-7** Heat-transfer data.

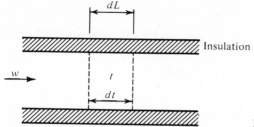

**Figure 2-8** Differential length of pipe.

area $A_i$ is

$$\frac{1}{UA_i} = \frac{1}{113A_i} + \frac{x}{k[(A_i + A_o)/2]} + \frac{1}{7.2A_o}$$

The temperature change $dt$ in a differential length of pipe (see Fig. 2-8) is expressed by

$$wc_p\,(-dt) = U\pi D[t - (-4°C)]\,dL$$

where $w$ = mass rate of flow = 44 kg/s

$c_p$ = specific heat of oil = 1930 J/(kg · K)

$t$ = oil temperature, °C

$dt$ = change in oil temperature in length $dL$

$U$ = overall heat-transfer coefficient between oil and permafrost, W/(m² · K)

$D$ = pipe diameter = 0.6 m

$L$ = length of pipe, m

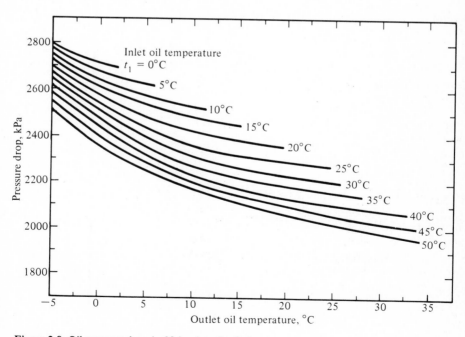

**Figure 2-9** Oil pressure drop in 32-km length of pipe.

The pressure drop in the 32-km section of pipe is a function of the inlet and outlet temperatures of the oil because of the influence of these temperatures on viscosity. Figure 2-9 shows pressure drops for 32 km. The maximum pressure the pipe can withstand is 2350 kPa gauge. Specify the following:

Insulation thickness _____ mm
Inlet oil temperature $t_1$ _____°C
Outlet oil temperature $t_2$ _____°C
Pressure drop _____ kPa
Temperature of surface in contact with permafrost (highest in 32-km run) _____°C

(*c*) Review the design and list the decisions that preclude possible optimization later in the design.

**2.5**[†] The tube spacing and fin height are to be selected for the steam-generating section of a furnace, shown in Fig. 2-10. The furnace section is 1.8 m wide; the tubes are 2.5 m long and are arranged in a square array, six rows high. The tubes have an OD of 75 mm, and they can be either bare or equipped with fins. The fins are 2 mm thick and are spaced 6 mm apart along the length of the tube.

The steam temperature in the boiler is 175°C, and the entering temperature of the stack gases is 560°C. The minimum spacing between the centerlines of the tubes, which is dictated by the smallest U bend available, is 125 mm. There are two restrictions on the design: (1) there must be 3 mm clearance between the fins of adjacent tubes and between the fins and the walls and (2) the maximum metal temperature, which occurs at the tip of the fins, must not exceed 415°C in order to limit oxidation.

[†]Problem based on a suggestion of O. B. Taliaferro, Exxon Research and Engineering Company.

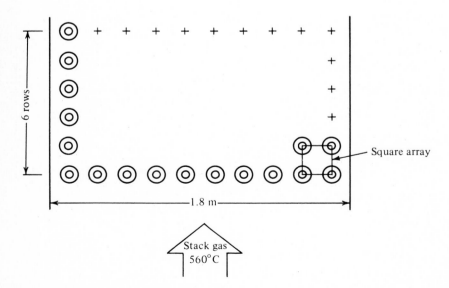

**Figure 2-10** Steam-generating section in a furnace.

**Table 2-3 Fin data**

| Fin height, mm, measured from outside of tube | $A_{fin}$, m²/m | $\eta$ | $\dfrac{560 - t_{tip}}{560 - t_{root}}$ |
|---|---|---|---|
| 0 | 0.079 | 1.00 | 1.00 |
| 12 | 1.199 | 0.94 | 0.88 |
| 18 | 1.872 | 0.86 | 0.77 |
| 25 | 2.754 | 0.75 | 0.67 |
| 31 | 3.591 | 0.64 | 0.58 |
| 37 | 4.503 | 0.55 | 0.49 |
| 44 | 5.665 | 0.46 | 0.41 |

The essential requirement of the furnace is that the product of $UA$ for the boiler section be 18,500 W/K or higher. The $U$ value, based on steam-side area, is

$$\frac{1}{U} = R_{total} = R_{steam} + R_{tube} + R_{gas}$$

where the resistances are

$$R_{steam} = \frac{1}{5700} = 0.0001754 \ (m^2 \cdot K)/W$$

$$R_{tube} = \frac{0.004 \text{ m thick}}{k = 44} = 0.0000909 \ (m^2 \cdot K)/W$$

$$R_{gas} = \frac{1}{h_g(A_{prime}/A_{steam}) + h_g(A_{fin}\,\eta/A_{steam})}$$

where $h_g$ = gas-side heat-transfer coefficient = 70 W/(m² · K)
$A_{prime}$ = 0.157 m² per meter of tube length
$A_{steam}$ = 0.210 m² per meter of tube length
$A_{fin}$ = fin area per meter of tube length
$\eta$ = fin effectiveness

Table 2-3 gives fin areas per meter of tube length, the fin effectiveness, and values of the expression for the fin-tip temperature as a function of the root temperature. In turn the root temperature can be computed from

$$\frac{t_{root} - 175}{560 - 175} = \frac{R_{steam} + R_{tube}}{R_{total}}$$

Specify the number of tubes in each horizontal row and the fin height so that the assembly fits into the space available (abiding by the necessary clearances), provides a $UA$ of 18,500 W/K or more, and maintains a fin-tip temperature of 415°C or less.

# REFERENCE

1. D. F. Othmer and J. W. E. Griemsmann, "Moving the Arctic Oil: Pipelines and the Pour Point," *Mech. Eng.*, vol. 93, no. 11, pp. 27–32, November 1971, describes further approaches to moving oil under Arctic conditions.

# THREE

## ECONOMICS

## 3-1 INTRODUCTION

The basis of most engineering decisions is economic. Designing and building a device or system that functions properly is only part of the engineer's task. The device or system must, in addition, be economic, which means that the investment must show an adequate return. In the study of thermal systems, one of the key ingredients is optimization, and the function that is most frequently optimized is the potential profit. Sometimes the designer seeks the solution having minimum first cost or, more frequently, the minimum total lifetime cost of the facility.

Hardly ever are decisions made solely on the basis of monetary considerations. Many noneconomic factors affect the decisions of industrial organizations. Decisions are often influenced by legal concerns, such as zoning regulations, or by social concerns, such as the displacement of workers, or by air or stream pollution. Aesthetics also have their influence, e.g., when extra money is spent to make a new factory building attractive. Since these social or aesthetic concerns almost always require the outlay of extra money, they revert to such economic questions as how much a firm is willing or able to spend for locating a plant where the employees will live in a district with good schools.

This chapter first explains the practice of charging interest and then proceeds to the application of interest in evaluating the worth of lump sums, of series of uniform payments, and of payments that vary linearly with time. Numerous applications of these factors will be explored, including such standard and important ones as computing the value of bonds. Methods of making economic comparisons of alternatives, the influence of taxes, several methods of computing depreciation, and continuous compounding will be explained.

## 3-2 INTEREST

Interest is the rental charge for the use of money. When renting a house, a tenant pays rent but also returns possession of the house to the owner after the stipulated period. In a simple loan, the borrower of money pays the interest at stated periods throughout the duration of the loan, e.g., every 6 months or every year, and then returns the original sum to the lender.

The existence of interest gives money a time value. Because of interest it is not adequate simply to total all the expected lifetime receipts and in another column total all the expected lifetime expenditures of a facility and subtract the latter from the former to determine the profit. A dollar at year 4 does not have the same value as a dollar at year 8 (even neglecting possible inflation) due to the existence of interest. A thought process that must become ingrained in anyone making economic calculations is that the worth of money has two dimensions, the dollar amount and the time. Because of this extra dimension of time, equations structured for solution of economic problems must equate amounts that are all referred to a common time base.

The most fundamental type of interest is *simple interest*, which will be quickly dismissed because it is hardly ever applied.

**Example 3-1** Simple interest of 8 percent per year is charged on a 5-year loan of $500. How much does the borrower pay to the lender?

SOLUTION The annual interest is ($500)(0.08) = $40, so at the end of 5 years the borrower pays back to the lender $500 + 5($40) = $700.

## 3-3 LUMP SUM, COMPOUNDED ANNUALLY

Annual compounding means that the interest on the principal becomes available at the end of each year, this interest is added to the principal, and the combination draws interest during the next year. In Example 3-1 if the interest were compounded annually, the value at the end of the first year would be

$$\$500 + (\$500)(0.08) = \$540$$

**Table 3-1**

| Year | Interest during year | Amount $S$ at end of year |
|------|---------------------|---------------------------|
| 1 | $Pi$ | $P + Pi = P(1 + i)$ |
| 2 | $P(1 + i)i$ | $P(1 + i) + P(1 + i)i = P(1 + i)^2$ |
| . . . . . . | . . . . . . . . . . . . . | . . . . . . . . . . . . . . . . . . . . . . . . . . . . . . . |
| $n$ | $P(1 + i)^{n-1}i$ | $P(1 + i)^{n-1} + P(1 + i)^{n-1}i = P(1 + i)^n$ |

During the second year the interest is computed on $540, so that at the end of the second year the value is

$$\$540 + (\$540)(0.08) = \$583.20$$

At the end of the third year the value is

$$\$583.20 + (\$583.20)(0.08) = \$629.86$$

The pattern for computing the value of an original amount $P$ subjected to interest compounded annually at a rate $i$ is shown in Table 3-1.

**Example 3-2** What amount must be repaid on the $500 loan in Example 3-1 if the interest of 8 percent is compounded annually?

SOLUTION Amount to be repaid after 5 years $= (\$500)(1 + 0.08)^5 = \$734.66$.

# 3-4 LUMP SUM COMPOUNDED MORE OFTEN THAN ANNUALLY

In most business situations compounding is often semiannually, quarterly, or even daily. At compounding periods other than annual a special interpretation is placed on the designation of the interest rate in that $i$ refers to *nominal annual interest rate*. If interest is compounded semiannually, an interest of $i/2$ is assessed every half year.

**Example 3-3** What amount must be repaid on a 5-year $500 loan at 8 percent annual interest compounded quarterly?

SOLUTION

$$\text{Amount to be repaid} = (\$500)\left(1 + \frac{0.08}{4}\right)^{20} = \$742.97$$

The interpretation of the 8 percent interest in Example 3-3 is that one-fourth of that rate is assessed each quarter and there are 20 quarters, or compounding periods, in the 5-year span of the loan.

## 3-5  COMPOUND-AMOUNT FACTOR (f/p) AND PRESENT-WORTH FACTOR (p/f)

The factors that translate the value of lump sums between present and future worths are

Compound-amount factor (CAF or f/p)
Present-worth factor (PWF or p/f)

The application of these factors is

$$\text{Future worth} = (\text{present worth})(f/p)$$
$$\text{Present worth} = (\text{future worth})(p/f)$$

The two factors are the reciprocal of each other, and the expressions for the factors are

$$f/p = \left(1 + \frac{i}{m}\right)^{mn} \quad \text{and} \quad p/f = \frac{1}{(1 + i/m)^{mn}}$$

where $i$ = nominal annual interest rate
$n$ = number of years
$m$ = number of compounding periods per year

**Example 3-4**  You invest $5000 in a credit union which compounds 5 percent interest quarterly. What is the value of the investment after 5 years?

SOLUTION

$$\text{Future amount} = (\text{present amount})\left(f/p, \frac{0.05}{4}, 20 \text{ periods}\right)$$

where the meaning of the convention is (factor, rate, period).

$$\text{Future amount} = (\$5000)\left(1 + \frac{0.05}{4}\right)^{20} = (\$5000)(1.2820) = \$6410$$

**Example 3-5**  A family wishes to invest a sum of money when a child begins elementary school so that the accumulated amount will be $10,000 when the child begins college 12 years later. The money can be invested at 8 percent, compounded semiannually. What amount must be invested?

SOLUTION

$$\text{Present amount} = (\text{future amount})\left(p/f, \frac{0.08}{2}, 24 \text{ periods}\right)$$

$$= \frac{\$10,000}{(1 + 0.08/2)^{24}} = \frac{10,000}{2.5633} = \$3901.20$$

## 3-6 FUTURE WORTH (f/a) OF A UNIFORM SERIES OF AMOUNTS

The next factor to be developed translates the value of a uniform series of amounts into the value some time in the future. The equivalence is shown symbolically in Fig. 3-1, where $R$ is the uniform amount at each time period. The arrow indicating future worth is the equivalent value in the future of that series of regular amounts with the appropriate interest applied.

Suppose that in Fig. 3-1 the magnitude of $R$ is \$100 and that this amount appears on an annual basis. Further assume that the interest rate is 6 percent, compounded annually. At the end of the first year the accumulated sum $S$ is the \$100 that has just become available. At the end of 2 years the accumulated sum is the \$100 from the first year with its interest plus the new \$100 amount:

$$S = (\$100)(1 + 0.06) + \$100 = (\$100)[(1 + 0.06) + 1]$$

At the end of 7 years

$$S = (\$100)[(1 + 0.06)^6 + (1 + 0.06)^5 + \cdots + (1 + 0.06) + 1]$$
$$= (\$100)(8.3938) = \$839.38$$

In general, the future worth $S$ of a uniform series of amounts, each of which is $R$, with an interest rate $i$ that is compounded at the same frequency as the $R$ amounts is

$$S = R[(1 + i)^{n-1} + (1 + i)^{n-2} + \cdots + (1 + i) + 1]$$

The term in the brackets is called the *series-compound-amount factor* (SCAF or f/a)

$$f/a = (1 + i)^{n-1} + (1 + i)^{n-2} + \cdots + (1 + i) + 1 \tag{3-1}$$

A closed form for the series can be developed by first multiplying both sides of Eq. (3-1) by $1 + i$

$$(f/a)(1 + i) = (1 + i)^n + (1 + i)^{n-1} + \cdots + (1 + i) \tag{3-2}$$

Substracting Eq. (3-1) from Eq. (3-2) yields

$$(f/a)[(1 + i) - 1] = (1 + i)^n - 1$$

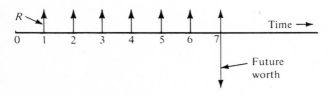

Figure 3-1 Future worth of a series of amounts.

and so

$$f/a = \frac{(1 + i)^n - 1}{i} \qquad (3\text{-}3)$$

The inverse of f/a, called the sinking-fund factor (SFF or a/f), is the multiplying factor that provides the regular amount $R$ when the future worth is known:

Regular amount $R$ = (future worth) (SFF) = (future worth) (a/f)

The term a/f is the reciprocal of f/a:

$$a/f = \frac{i}{(1 + i)^n - 1} \qquad (3\text{-}4)$$

A point to emphasize is the convention for when the first regular amount becomes available. Figure 3-1 shows that the first amount $R$ is available at the end of the first period and not at time 0. Sometimes when a series is to be evaluated the first amount may be available at time 0; such conversions can be made and will be explained later in the chapter. The important point is that whenever Eq. (3-3) or (3-4) is used, it applies to the convention by which the first of the regular amounts occurs at the end of the first time period.

**Example 3-6** A new machine has just been installed in a factory, and the management wants to set aside and invest equal amounts each year starting 1 year from now so that $16,000 will be available in 10 years for the replacement of the machine. The interest received on the money invested is 8 percent, compounded annually. How much must be provided each year?

SOLUTION  The annual amount $R$ is

$$R = (\$16{,}000)(a/f, 8\%, 10) = (\$16{,}000)\frac{0.08}{(1 + 0.08)^{10} - 1}$$

$$= (\$16{,}000)(0.06903) = \$1104.50$$

## 3-7 PRESENT WORTH (p/a) OF A UNIFORM SERIES OF AMOUNTS

The *series-present-worth factor* (SPWF or p/a) translates the value of a series of uniform amounts $R$ into the present worth, as shown symbolically in Fig. 3-2. As in the previous section, the convention is that the first payment occurs at the end of the first time period and not at time 0. The present worth of the series can be found by applying the p/f factor to each of the $R$ amounts:

$$\text{Present worth} = \frac{R}{(1 + i)^1} + \frac{R}{(1 + i)^2} + \cdots + \frac{R}{(1 + i)^n}$$

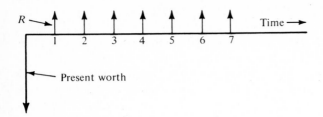

**Figure 3-2** Present worth of a series of amounts.

or

$$\text{Present worth} = R \left[ \frac{1}{(1+i)^1} + \frac{1}{(1+i)^2} + \cdots + \frac{1}{(1+i)^n} \right] \quad (3\text{-}5)$$

If both sides of Eq. (3-5) are multiplied by $(1+i)^n$, the expression in the brackets after multiplying by $(1+i)^n$ is the same as f/a in Eq. (3-1). Thus

$$(\text{Present worth}) (1+i)^n = R(\text{f/a}) = R \frac{(1+i)^n - 1}{i}$$

The series-present-worth factor p/a is the one which when multiplied by $R$ yields the present worth. Thus

$$\text{p/a} = \frac{(1+i)^n - 1}{i(1+i)^n} \quad (3\text{-}6)$$

It is not surprising to find from a comparison of Eqs. (3-3) and (3-6) that p/a is (f/a) $[1/(1+i)^n]$. Thus

$$\text{p/a} = (\text{f/a}) (\text{p/f})$$

The reciprocal of the series-present-worth factor a/p is called the *capital-recovery factor*.

$$\text{a/p} = \frac{i(1+i)^n}{(1+i)^n - 1} \quad (3\text{-}7)$$

**Example 3-7** You borrow $1000 from a loan company that charges 15 percent nominal annual interest compounded monthly. You can afford to pay off $38 per month on the loan. How many months will it take to repay the loan?

SOLUTION

$$\$1000 = (\$38) (\text{p/a}, 1.25\%, n \text{ months})$$

$$1000 = (38) \frac{(1.0125)^n - 1}{(0.0125)(1.0125)^n}$$

$$(1.0125)^n = 1.490$$

$$n = 32.1 \text{ months}$$

## 3-8 GRADIENT-PRESENT-WORTH FACTOR

The factors p/a, f/a, and their reciprocals apply where the amounts in the series are *uniform*. Instead the present worth of a series of *increasing* amounts may be sought. Typical cases that the series of increasing amounts approximates are maintenance costs and energy costs. The cost of maintenance of equipment is expected to increase progressively as the equipment ages, and energy costs may be projected to increase as fossil fuels become more expensive.

The gradient-present-worth factor (GPWF) applies when there is no cost during the first year, a cost $G$ at the end of the second year, $2G$ at the end of the third year, and so on, as shown graphically in Fig. 3-3. The straight line in Fig. 3-3 starts at the end of year 1. The present worth of this series of increasing amounts is the sum of the individual present worths

$$\text{Present worth} = \frac{G}{(1+i)^2} + \frac{2G}{(1+i)^3} + \cdots + \frac{(n-1)G}{(1+i)^n}$$

This series can be expressed in closed form as

$$\text{Present worth} = G(\text{GPWF}) = G\left\{\frac{1}{i}\left[\frac{(1+i)^n - 1}{i(1+i)^n} - \frac{n}{(1+i)^n}\right]\right\}$$

**Example 3-8** The annual cost of energy for a facility is $3000 for the first year (assume payable at the end of the year) and increases by 10 percent, or $300, each year thereafter. What is the present worth of this 12-year series of energy costs, as shown in Fig. 3-4, if the interest rate is 9 percent compounded annually?

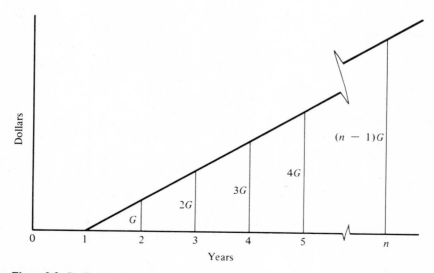

**Figure 3-3** Gradient series.

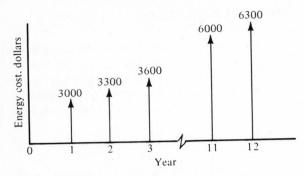

**Figure 3-4** Progressively increasing energy costs in Example 3-8.

SOLUTION The series of amounts shown in Fig. 3-4 could be reproduced by a combination of two series, a uniform series of $3000 and a gradient series in which $G = 300$. Thus

Present worth = (3000) (p/a, 9%, 12) + (300) (GPWF, 9%, 12)

$$PW = (\$3000) \frac{(1.09)^{12} - 1}{(0.09)(1.09)^{12}} + (\$300)\left\{\frac{1}{0.09}\left[\frac{(1.09)^{12} - 1}{(0.09)(1.09)^{12}} - \frac{12}{(1.09)^{12}}\right]\right\}$$

$PW = \$21,482 + \$9648 = \$31,130$

## 3-9 SUMMARY OF INTEREST FACTORS

The factors p/f, f/a, p/a, their reciprocals, and the GPWF are tools that can be judiciously applied and combined to solve a variety of economic problems. These factors are summarized in Table 3-2.

**Table 3-2 Summary of interest factors**

| Factor | Formula |
|--------|---------|
| f/p | $(1 + i)^n$ |
| f/a | $\dfrac{(1 + i)^n - 1}{i}$ |
| p/a | $\dfrac{(1 + i)^n - 1}{i(1 + i)^n}$ |
| GPWF | $\dfrac{1}{i}\left[\dfrac{(1 + i)^n - 1}{i(1 + i)^n} - \dfrac{n}{(1 + i)^n}\right]$ |

Sections 3-10 to 3-13 and a number of problems at the end of the chapter illustrate how these factors can be combined to solve more complicated situations. In most of these problems the solution depends on setting up an equation that expresses the equivalence of amounts existing at different times. Many of these problems are thus solved by translating amounts to the same basis, such as the future worth of all amounts, the present worth, or the annual worth. One point to emphasize is that the influence of interest means that a given sum has differing values at different times. As a consequence the equations just mentioned should never add or subtract amounts applicable to different times. The amounts must always be translated to a common base first.

One of the important financial transactions is the issuance, selling, and buying of bonds. Bonds will be the first application of the combination of factors shown in Table 3-2.

## 3-10 BONDS

Commercial firms have various methods for raising capital to conduct or expand their business. They may borrow money from banks, sell common stock, sell preferred stock, or issue bonds. A bond is usually issued with a face value of $1000, which means that at maturity the bond's owner surrenders it to the firm that issued it and receives $1000 in return. In addition, interest is paid by the firm to the owner at a rate specified on the bond, usually semiannually. The owner of a bond therefore receives $(i/2)(\$1000)$ every 6 months plus $1000 at maturity. The original purchaser does not have to keep the bond for the life of the bond but may sell it at a price agreed on with a purchaser. The price may be higher or lower than $1000.

Interest rates fluctuate, depending upon the availability of investment money. A firm may have to pay 10 percent interest in order to sell the bond at the time it is issued, but 2 years later, for example, the going rate of interest may be above or below 10 percent. The owner of a 10 percent bond can demand more than $1000 for the bond if the going interest rate is, say, 8 percent, but would have to sell the bond for less than $1000 if the going rate of interest were 12 percent. In other words, the current price of the bond is such that the total investment earns the current rate of interest.

The principle of setting up an equation that reflects all values to a common basis will now be applied by equating what should be paid for the bond to what is acquired from the bond. Arbitrarily choose the future worth as the basis of all amounts

$$P_b \left( f/p, \frac{i_c}{2}, 2n \right) = \left( f/a, \frac{i_c}{2}, 2n \right) \frac{i_b}{2} (1000) + 1000 \qquad (3-8)$$

where $P_b$ = price to be paid for bond now
$i_c$ = current rate of interest

$i_b$ = rate of interest on bond
$n$ = years to maturity

The terms in Eq. (3-8) are as follows. On the left side of the equation is the future worth of the investment, which is the price to be paid for the bond translated to the future. The interest applicable is the current rate of interest, which is assumed to continue from now until maturity of the bond.

The first term on the right of Eq. (3-8) is the future worth of a uniform series of the semiannual interest payments on the bond. It is assumed that the investor immediately reinvests the interest payments at the going rate of interest $i_c$. The other term on the right of the equation is $1000, which is the amount the firm will pay back to the owner of the bond at maturity. This $1000 is already at the future time, so no correction need be made on it.

**Example 3-9** A $1000 bond that has 10 years to maturity pays interest semiannually at a nominal annual rate of 8 percent. An investor wishes to earn 9 percent on her investment. What price could she pay for the bond in order to achieve this 9 percent interest rate?

SOLUTION  Application of Eq. (3-8) gives

$$P_b(f/p, 0.045, 20) = (f/a, 0.045, 20) (0.04) (1000) + 1000$$

$$P_b(2.4117) = (31.371) (0.04) (1000) + 1000$$

$$P_b = \$934.96$$

## 3-11  SHIFT IN TIME OF A SERIES

In the series shown in Fig. 3-1 for future worth and Fig. 3-2 for present worth, the convention is that the first regular amount appears at the end and not the beginning of the first period. In an actual situation the first amount may appear at time 0, and no amount appears at the end, as illustrated in Fig. 3-5. If the future worth of this series is sought, for example, Eq. (3-1) would be modified by multiplying each of the terms by $1 + i$. The closed form in Eq. (3-3) could be multiplied by $1 + i$, with the result that the f/a for the series in Fig. 3-5 is

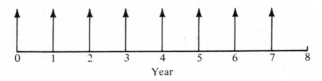

**Figure 3-5** First amount in a series appearing at time 0.

$$(f/a)_{shift} = (1 + i) \, \frac{(1 + i)^n - 1}{i} \qquad (3\text{-}9)$$

## 3-12 DIFFERENT FREQUENCY OF SERIES AMOUNTS AND COMPOUNDING

The formulas for f/a and p/a were developed on the basis of identical compounding and payment frequencies, e.g., annual or semiannual. A situation often arises where the periods are different, e.g., quarterly compounding with annual series amounts. More frequent compounding than payments can be accommodated by determining an equivalent rate of interest applicable between the payment periods.

Example 3-10 Annual investments of $1200 are to be made at a savings and loan institution for 10 years beginning at the end of the first year. The institution compounds interest quarterly at a nominal annual rate of 5 percent. What is the expected value of the investment at the end of 10 years?

SOLUTION If an amount $X$ starts drawing interest at the beginning of a year and is compounded quarterly at an annual rate of 5 percent, the value $Y$ at the end of the year is

$$Y = X\left(1 + \frac{0.05}{4}\right)^4 = X(1.0509) = X(1 + 0.0509)$$

The equivalent annual rate of interest due to the quarterly compounding is therefore 5.09 percent. This rate can be used in the f/a factor for the annual amounts

Future worth = (f/a, 5.09%, 10 years) ($1200)

$$= \frac{(1.0509)^{10} - 1}{0.0509} (1200) = (12.631)(1200) = \$15,157$$

## 3-13 CHANGES IN MIDSTREAM

A plan involving a series of payments (or withdrawals) is set up on the basis of certain regular amounts and expected interest rates. During the life of the plan the accepted interest rate may change and/or the ability to make the regular payments may change, and an alteration in the plan may be desired. This class of problems can generally be solved by establishing the worth at the time of change and making new calculations for the remaining term.

**Example 3-11** A sinking fund is established such that $12,000 will be available to replace a facility at the end of 10 years. At the end of 4 years, following the fourth uniform payment, management decides to retire the facility at the end of 9 years of life. At an interest rate of 6 percent, compounded annually, what are the payments during the first 4 and last 5 years?

SOLUTION According to the original plan, the payments to be made for 10 years were

$$(\$12{,}000)\,(a/f,\,6\%,\,10) = (12{,}000)\,(0.075868) = \$910.42$$

The worth of the series of payments at the end of 4 years is

$$(\$910.42)\,(f/a,\,6\%,\,4) = (910.42)\,(4.3746) = \$3982.72$$

This accumulated value will draw interest for the next 5 years, but the remainder of the $12,000 must be provided by the five additional payments plus interest

$$R(f/a,\,6\%,\,5) = \$12{,}000 - (3982.72)\,(f/p,\,6\%,\,5)$$

and so the payments during the last 5 years must be

$$R = \frac{12{,}000 - (3{,}982.72)\,(1.3382)}{5.6371} = \$1183.30$$

## 3-14 EVALUATING POTENTIAL INVESTMENTS

An important function of economic analyses in engineering enterprises is to evaluate proposed investments. A commercial firm must develop a rate of return on its investment that is sufficient to pay corporation taxes and still leave enough to pay interest on the bonds or dividends on the stock that provide the investment capital. The evaluations can become very intricate, and only the basic investment situations will be explained. This fundamental approach is, however, the starting point from which modifications and refinements can be made in more complicated situations.

Four elements will be considered in investment analyses: (1) first cost, (2) income, (3) operating expense, and (4) salvage value. The *rate of return* is treated as though it were interest.

**Example 3-12** An owner-manager firm of rental office buildings has a choice of buying building $A$ or building $B$ with the intent of operating the building for 5 years and then selling it. Building $A$ is in an improving location, so that the expected value is to be 20 percent higher in 5 years, while building $B$ is in a declining neighborhood, with an expected drop in value of 10 percent in 5 years. Other data applicable to these buildings are shown in Table 3-3. What will be the rate of return on each building?

**Table 3-3 Economic data for Example 3-12**

|  | Building $A$ | Building $B$ |
|---|---|---|
| First cost | $800,000 | $600,000 |
| Annual income from rent | 160,000 | 155,000 |
| Annual operating and maintenance cost | 73,000 | 50,300 |
| Anticipated selling price | 960,000 | 540,000 |

SOLUTION The various quantities must be translated to a common basis, future worth, annual worth, or present worth. If all amounts are placed on a present-worth basis, the following equations apply:

Building $A$:

$$800,000 = (160,000 - 73,000) \, (p/a, i\%, 5) + (960,000) \, (p/f, i\%, 5)$$

Building $B$:

$$600,000 = (155,000 - 50,300) \, (p/a, i\%, 5) + (540,000) \, (p/f, i\%, 5)$$

The unknown in each of the two present-worth equations is the rate of return $i$. Since the expressions are not linear in $i$, an iterative technique must be used to find the value of $i$

$$i = \begin{cases} 13.9\% & \text{building } A \\ 16.0\% & \text{building } B \end{cases}$$

Building $B$ provides a greater rate of return, and the firm must now decide whether even the 16 percent is adequate to justify the investment.

# 3-15 TAXES

The money for operating the government and for financing services provided by the government comes primarily from taxes. The inclusion of taxes in an economic analysis is often important because in some cases taxes may be the factor deciding whether to undertake the project or not. In certain other cases the introduction of tax considerations may influence which of two alternatives will be more attractive economically.

In most sections of the United States, property taxes are levied by a sub-state taxing district in order to pay for schools, city government and services, and perhaps park and sewage systems. Theoretically, the real estate tax should decrease as the facility depreciates, resulting in lower real estate taxes as the facility ages. Often on investments such as buildings, the tax, as a dollar figure, never decreases. It is therefore a common practice to plan for a constant real estate tax when making the investment analysis. The effect of the tax is to penalize a facility which has a high taxable value.

Federal corporation income taxes on any but the smallest enterprises amount to approximately 50 percent of the profit. In Example 3-12 the rate of return on the investment in building $B$ was 16 percent, which may seem very favorable compared with the usual range of interest rates of 5 to 10 percent. The rate of return of 16 percent is before taxes, however; after the corporation income tax has been extracted, the rate of return available for stock or bond holders in the company is of the order of 8 percent. Since income tax is usually a much more significant factor in the economic analysis than property tax, income tax will be discussed further. An ingredient of income tax calculations is depreciation, explored in the next section.

## 3-16  DEPRECIATION

Depreciation is an amount that is listed as an annual expense in the tax calculation to allow for replacement of the facility at the end of its life. Numerous methods of computing depreciation are permitted by the Internal Revenue Service, e.g., straight line, sum-of-the-year's digits, and double-rate declining balance. The first two will be explained.

Straight-line depreciation consists simply of dividing the difference between the first cost and salvage value of the facility by the number of years of tax life. The result is the annual depreciation. The tax life to be used is prescribed by the Internal Revenue Service and may or may not be the same as the economic life used in the economic analysis.

In the sum-of-the-year's-digits (SYD) method, the depreciation for a given year is represented by the formula

$$\text{Depreciation, dollars} = 2 \ \frac{N - t + 1}{N(N + 1)} \ (P - S) \tag{3-10}$$

where $N$ = tax life, years
$\quad\quad t$ = year in question
$\quad\quad P$ = first cost, dollars
$\quad\quad S$ = salvage value, dollars

If the tax life is 10 years, for example, the depreciation is

$$\text{First year} = 2(10/110) \ (P - S)$$
$$\text{Second year} = 2(9/110) \ (P - S)$$
$$\cdots\cdots\cdots\cdots\cdots\cdots\cdots\cdots\cdots\cdots\cdots$$
$$\text{Tenth year} = 2(1/110) \ (P - S)$$

A comparison of the depreciation rates calculated by the straight-line method with those calculated by the SYD method shows that the SYD method permits greater depreciation in the early portion of the life. With the SYD method the income tax that must be paid early in the life of the facility is less than with the straight-line method, although near the end of the life the SYD tax is greater. The total tax paid over the tax life of the facility is the same by either method, but the advantage of using the SYD method is that more of the tax is paid in later years, which is advantageous in view of the time value of money.

The straight-line method has an advantage, however, if it is likely the tax rate will increase. If the rate jumps, it is better to have paid the low tax on a larger fraction of the investment.

## 3-17 INFLUENCE OF INCOME TAX ON ECONOMIC ANALYSIS

To see the effect of depreciation and federal income tax, consider the following simple example of choosing between alternative investments $A$ and $B$, for which the data in Table 3-4 apply.

A calculation of the annual cost of both alternatives without inclusion of the income tax is as follows:

| | |
|---|---:|
| Alternative $A$: | |
| First cost on annual basis (200,000) (a/p, 9%, 20) | $21,910 |
| Annual operating expense | 14,000 |
| Real estate tax and insurance | 10,000 |
| Total | $45,910 |
| Alternative $B$: | |
| First cost on annual basis (270,000) (a/p, 9%, 30) | $26,280 |
| Annual operating expense | 6,200 |
| Real estate tax and insurance | 13,500 |
| Total | $45,980 |

## Table 3-4 Alternative investments

| | Alternative $A$ | Alternative $B$ |
|---|---|---|
| First cost | $200,000 | $270,000 |
| Life | 20 years | 30 years |
| Salvage value | 0 | 0 |
| Annual income | $60,000 | $60,000 |
| Annual operating expense | $14,000 | $6,200 |
| Real estate tax and insurance (5% of first cost) | $10,000 | $13,500 |
| Interest | 9% | 9% |

**Table 3-5 Income tax on two alternatives**

| First-year expenses | Alternative $A$ | Alternative $B$ |
|---|---|---|
| Depreciation | $10,000 | $ 9,000 |
| Interest, 9% of unpaid balance | 18,000 | 24,300 |
| Operating expense, tax and insurance | 24,000 | 19,700 |
| Total expenses | $52,000 | $53,000 |
| Profit = income – expenses | 8,000 | 7,000 |
| Income tax (50% of profits) | 4,000 | 3,500 |

The economic analysis of alternatives $A$ and $B$ shows approximately the same annual costs and incomes (in fact, the example was rigged to accomplish this).

In the computation of profit on which to pay income tax, the actual interest paid is listed as an expense, and if straight-line depreciation is applied, the expenses for the first year for the two alternatives are as shown in Table 3-5.

A higher income tax must be paid on $A$ than on $B$ during the early years. In the later years of the project, a higher tax will be paid on $B$. The example shows that even though the investment analysis indicated equal profit on the two alternatives, the inclusion of income tax shifts the preference to $B$. The advantage of $B$ is that the present worth of the tax payments is less than for $A$.

## 3-18 CONTINUOUS COMPOUNDING

Frequencies of compounding such as annual, semiannual, and quarterly have been discussed. Even shorter compounding periods are common, e.g., the daily compounding offered investors by many savings and loan institutions. High frequency of compounding is quite realistic in business operation, because the notion of accumulating money for a quarterly or semiannual payment is not a typical practice. Businesses control their money more on a flow basis than on a batch basis. The limit of compounding frequency is *continuous compounding* with an infinite number of compounding periods per year. This section discusses three factors applicable to continuous compounding: (1) the continuous compounding factor corresponding to f/p, (2) uniform lump sums continuously compounded, and (3) continuous flow continuously compounded.

The f/p term with a nominal annual interest rate of $i$ compounded $m$ times per year for a period of $n$ years is

$$f/p = \left(1 + \frac{i}{m}\right)^{mn}$$

The f/p factor for continuous compounding $(f/p)_{cont}$ is found by letting $m$ approach infinity

$$(f/p)_{cont} = \left(1 + \frac{i}{m}\right)^{mn}\Bigg|_{m \to \infty} \tag{3-11}$$

To evaluate the limit, take the logarithm of both sides

$$\ln (f/p)_{cont} = mn \left[ \ln \left( 1 + \frac{i}{m} \right) \right]_{m \to \infty}$$

Express $\ln (1 + i/m)$ as an infinite series of $i/m$

$$\ln (f/p)_{cont} = mn \left[ 0 + \frac{i}{m} + (const) \frac{i^2}{m^2} + \cdots \right]_{m \to \infty}$$

Cancel $m$ and let $m \to \infty$

$$\ln (f/p)_{cont} = in$$

Therefore

$$(f/p)_{cont} = e^{in} \tag{3-12}$$

**Example 3-13**  Compare the values of $(f/p, 8\%, 10)$ and $[(f/p)_{cont}, 8\%, 10]$.

SOLUTION

$$(f/p, 8\%, 10) = (1 + 0.08)^{10} = 2.1589$$

$$[(f/p)_{cont}, 8\%, 10] = e^{0.8} = 2.2255$$

The next factor presented is that of the future worth of a uniform series of lumped amounts compounded continuously. The continuous f/a factor for annual amounts $R$ can be derived by modifying Eq. (3-1) to continuous compounding by replacing $1 + i$ with $e^i$

$$(f/a)_{cont,lump} = (e^i)^{n-1} + (e^i)^{n-2} + \cdots + e^i + 1$$

In the closed-form expression of Eq. (3-3) $1 + i$ can be replaced by $e^i$ and $i$ by $e^i - 1$, yielding

$$(f/a)_{cont,lump} = \frac{e^{in} - 1}{e^i - 1} \tag{3-13}$$

The final continuous-compounding factor to be presented is the *continuous-flow future-worth factor* $(f/a)_{flow}$, which expresses the future worth of a continuous flow compounded continuously. If \$1 per year is divided into $m$ equal amounts and spread uniformly over the entire year, and if each of these $m$ amounts begins drawing interest immediately, the future worth of this series after $n$ years is

$$\frac{\$1}{m} \left( 1 + \frac{i}{m} \right)^{mn-1} + \frac{\$1}{m} \left( 1 + \frac{i}{m} \right)^{mn-2} + \cdots + \frac{\$1}{m} \left( 1 + \frac{i}{m} \right)^1 + \frac{\$1}{m} \tag{3-14}$$

If we use the analogy between Eqs. (3-1) and (3-3), Eq. (3-14) can be translated into the closed form

$$(f/a)_{flow} \text{ for } \$1 \text{ per year} = \frac{1}{m} \frac{(1 + i/m)^{mn} - 1}{i/m} = \frac{(1 + i/m)^{mn} - 1}{i}$$

As $m$ approaches infinity, the term $(1 + i/m)^{mn}$ approaches $e^{in}$, so that

$$(f/a)_{flow} \text{ for \$1 per year} = \frac{e^{in} - 1}{i} \qquad (3\text{-}15)$$

**Example 3-14** Location $A$ for a factory is expected to produce an annual profit that is \$10,000 per year greater than if location $B$ is chosen. This additional profit is spread evenly over the entire year and constitutes a continuous flow of \$10,000 per year. The profit is continuously reinvested at the rate of return of 14 percent per year. At the end of 8 years, the expected economic life of the factories, what is the difference between the worths of the two investments?

SOLUTION Since the \$10,000 per year difference is a continuous flow continuously compounded, the difference between future worths is

$$(\$10,000) \frac{e^{(0.14)(8)} - 1}{0.14} = \$147,490$$

## 3-19  SUMMARY

There are several levels of economic analysis higher than that approached by this chapter. The complications in accounting, financing, and tax computations involve sophistications beyond those presented here. The stage achieved by this chapter might be described as the second level of economic analysis. The first level would be a trivial one of simply totaling costs with no consideration of the time value of money. The second level introduces the influences of interest, which imposes the dimension of time as well as amount in assessing the value of money.

The methods of investment analyses explained in this chapter are used repeatedly in engineering practice, and in most cases engineers are not required to go beyond these principles. These methods also are the base for extensions into more complex economic analyses.

## PROBLEMS

**3-1** Using a computer, calculate your personal set of tables for the factors f/p, f/a, p/a, and GPWF. Devote a separate page to each of the factors, label adequately, and calculate at the interest rates of 4, 5, 6, 7, 8, 9, 10, 11, 12, 14, 16, 18, 20, 25 percent. Calculate for the following interest periods: 1 to 20 by ones, 22 to 30 by twos, and 30 to 60 by fives. Print out the factors to four places after the decimal point.

**3-2** Annual investments are being made so that \$20,000 will be accumulated at the end of 10 years. The interest rate on these investments is initially expected to be 4 percent compounded annually. After 4 years, the rate of interest is unexpectedly increased to 5 percent, so that payments for the remaining 6 years can be reduced. What amounts should be invested annually for the first 4 years and what sums for the last 6?

Ans.: Final payment, $1547.

**3-3** A firm wishes to set aside equal amounts at the end of each of 10 years, beginning at the end of the first year, in order to have $8000 maintenance funds available at the end of the seventh, eighth, ninth, and tenth years. What is the required annual payment if the money is invested and draws 6 percent compounded annually?

Ans.: $2655.

**3-4** A home mortgage extends for 20 years at 8 percent interest compounded monthly. The payments are also made monthly. After how many months is half of the principal paid off?

Ans.: 164 months.

**3-5** A lender offers a 1-year loan at what he calls 8 percent interest but requires the interest to be paid at the beginning rather than at the end of the year, as the usual practice is. To what interest rate computed in the conventional manner does this interest charge correspond?

Ans.: 8.7%.

**3-6** A loan of $50,000 at 8 percent compounded annually is to be paid off in 25 years by uniform annual payments beginning at the end of the first year. These annual payments proceed on schedule until the end of the eighth year, when the borrower is unable to pay and misses the payment. He negotiates with the lender to increase the remaining 17 payments in such a way that the lender continues to receive 8 percent. What is the amount of the original and the final payments in the series?

Ans.: Final payments, $5197.44.

**3-7** An $18,000 mortgage on which 8 percent interest is paid, compounded monthly, is to be paid off in 15 years in equal monthly installments. What is the total amount of interest paid during the life of this mortgage?

Ans.: $12,964.

**3-8** What will be the future worth of a series of 15 annual $1000 payments if the nominal annual interest rate is 8 percent and the interest is compounded quarterly?

Ans.: $27,671.

**3-9** A sum of sufficient magnitude is to be invested now so that starting 10 years from now an amount of $2000 per year can be paid in each of 8 succeeding years. The unexpended money remains invested at 8 percent compounded annually. How much must be allocated now?

Ans.: $5749.50.

**3-10** A mortgage that was originally $20,000 is being paid off in regular quarterly payments of $500. The interest is 8 percent compounded quarterly. How much of the principal remains after 9 years, or 36 payments?

Ans.: $14,800.60.

**3-11** A 20-year mortgage set up for uniform monthly payments with 6 percent interest compounded monthly is taken over by a new owner after 8 years. At that time $12,000 is still owed on the principal. What was the amount of the original loan?

Ans.: $16,345.

**3-12** An investor buys common stock in a firm for $1000. At the end of the first year and every year thereafter, she receives a dividend of $100, which she immediately invests in a savings and loan institution that pays 5 percent interest compounded annually. At the end of the tenth year, just after receiving her dividend, she sells the stock for $1200. What is the rate of interest (on an annual compounding basis) yielded by this investment program?

Ans.: 9.41%.

**3-13** A sum of $20,000 is borrowed at an interest rate of 8 percent on the unpaid balance compounded semiannually. The loan is to be paid back with 10 equal payments in 20 years.

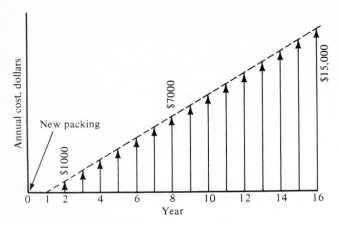

**Figure 3-6** Increasing costs due to cooling-tower deterioration.

The payments are to be made every 2 years, starting at the end of the second year. What is the amount of each biennial payment?

Ans.: $4291.

**3-14** The packing in a cooling tower that cools condensing water for a power plant progressively deteriorates and results in gradually rising costs due to reduced plant efficiency. These costs are treated as lump sums at the end of the year, as shown in Fig. 3-6. The cost is zero for the first year and then increases $1000 per year until the packing is 16 years old, when replacement is mandatory. At a point 8 years into the life of the packing (just after the $7000 annual cost has been assessed), a decision is to be made on the plan for the next 8 years, i.e., whether to replace the packing or to continue with the existing packing. Money can be borrowed at 9 percent interest, compounded annually. What is the maximum amount that could be paid for the packing in order to justify its replacement?

Ans.: $44,279.

**3-15** The anticipated taxes on a facility for its 10-year tax life decline in a straight-line fashion as follows:

| | |
|---|---|
| At end of year 1 | $10,000 |
| At end of year 2 | 9,000 |
| . . . . . . . . . . . . . . . . . . . . . | |
| At end of year 10 | 1,000 |

In an economic analysis of the facility the present worth of this series must be computed on the basis of 6 percent interest compounded annually.

(a) Using a combination of available factors, determine a formula for the present worth of a declining series like this one.

(b) Using the formula from part (a), compute the present worth of the above series.

Ans.: (b) $43,999.

**3-16** Calculate the uniform annual profit on a processing plant for which the following data apply:

| | |
|---|---|
| Life | 12 years |
| First cost | $280,000 |
| Annual real estate tax and insurance | 4% of first cost |
| Salvage value at end of 12 years | $50,000 |
| Annual cost of raw materials, labor, and other supplies | $60,000 |
| Annual income | $140,000 |
| Maintenance costs, during first year | 0 |
|     At end of second year | $1000 |
|     At end of third year | $2000 |
| .................................................... | |
|     At end of twelfth year | $11,000 |

The interest rate applicable is 6 percent compounded annually.
    **Ans.:** $33,560.

**3-17** A $1000 bond was issued 5 years ago and will mature 5 years from now. The bond yields an interest rate of 5 percent, or $50 per year. The owner of the bond wishes to sell the bond, but since interest rates have increased, a prospective buyer wishes to earn a rate of 6 percent on his investment. What should the selling price be? Remember that the purchaser receives $50 per year, which is reinvested, and receives the $1000 face value at maturity. Interest is compounded annually.
    **Ans.:** $957.88.

**3-18** Equation (3-8) relates the value of a bond $P_b$ to the bond interest and current rate of interest by reflecting all values to a future worth. Develop an equation that reflects all values to a uniform semiannual worth and solve Example 3-9 with this equation.

**3-19** Using a computer program, calculate tables of the price of a $1000 bond that will yield 5.0, 5.5, 6.0, 6.5, 7.0, 7.5, 8.0, 8.5, 9.0, 9.5, 10.0, 10.5, and 11.0 percent interest when the interest rates on the bond are 5.0, 5.5, 6.0, 6.5, 7.0, 7.5, 8.0, 8.5, 9.0, 9.5, and 10.0 percent. Compute the foregoing table for each of the following number of years to maturity: 2, 3, 4, 5, 6, 7, 8, 9, and 10. Interest is compounded semiannually.

**3-20** A municipality must build a new electric generating plant and can choose between a steam or a hydro facility. The anticipated cost of the steam plant is $10 million. Comparative data for the two plants are

| Plant | Generating cost, including maintenance, per kWh | Equipment life, years | Salvage value, percentage of first cost |
|---|---|---|---|
| Steam | $0.004 | 20 | 10 |
| Hydro | 0.002 | 30 | 10 |

The expected annual consumption of power is 300,000 MWh. If money is borrowed at 5 percent interest compounded annually, what first cost of the hydro plant would make the two alternatives equally attractive investments?
    **Ans.:** $21,593,000.

**3-21** A proposed investment consists of constructing a building, purchasing production machinery, and operating for 20 years. The expected life of the building is 20 years; its first cost is $250,000 with a salvage value of $50,000. Since the maximum life of the machinery is 12 years, it will be necessary to renew the machinery once during the 20 years. The first cost of the machinery is $132,000, and its salvage value is $132,000/(age, years). The annual income less the operating expense is expected to be $50,000. Annual interest is 6 percent compounded annually.

(a) When is the most favorable time to replace the machinery?

(b) Compute the present worth of the profit if the machinery is replaced at the time indicated by part (a).

Ans.: (b) $152,100.

**3-22** Owners of a plant that manufactures edible oil are considering constructing a tank to store unrefined oil; this will permit buying raw oil at more favorable prices. The cost of the tank is $150,000, it has an expected life of 10 years and a salvage value at the end of its life of $20,000. The anticipated annual saving in oil cost is $25,000. What is the rate of return on the investment?

Ans.: 11.65%.

**3-23** A new facility is expected to show zero profit during each of the first 5 years. For the 10 years thereafter the expected profit is to be $80,000 per year. If 12 percent interest (compounded annually) is desired as the return on the investment, what amount of investment is justified?

Ans.: $256,500.

**3-24** The anticipated income from an investment is $40,000 per year for the first 5 years and $30,000 per year for the remaining 5 years of life. The desired rate of return on the investment is 12 percent. The salvage value at the end of 10 years is expected to be 20 percent of the first cost. Determine the first cost that will result in the 12 percent return.

Ans.: $219,700.

**3-25** A processing plant has a first cost of $600,000 and an expected life of 15 years with no salvage value. Money is borrowed at 8 percent compounded annually, and the first cost is paid off with 15 equal annual payments. The expected annual income is $200,000, and annual operating expenses are $40,000. Corporation income tax is 50 percent of the profits before taxes, and the SYD method of depreciation is applicable on the tax life of the facility, which is 12 years with no salvage value. Compute the income tax for (a) the first year and (b) the second year.

Ans.: (a) $9846; (b) $14,576.

**3-26** A client who is constructing a warehouse instructs the contractor to omit insulation. The client explains that he will operate the building for several months and then install the insulation as a repair, so that he can deduct the expense from income tax at the end of the first year rather than spread it as straight-line depreciation over the 8-year tax life of the warehouse. The contractor points out that a later installation will cost more than the $20,000 cost of installing the insulation with the original construction. How much could the client afford to pay for the later installation for equal profit if he plans on a 15 percent return on his investment and corporation income taxes are 50 percent?

Ans.: $25,461.

**3-27** A $200,000 facility has an 8-year tax life, and the firm expects a 12 percent return on its investment and pays 50 percent corporation income tax on profits. The firm is comparing the relative advantage of the SYD and straight-line methods of depreciation. If the taxes computed by the two methods are expressed as uniform annual amounts, what is the advantage of the SYD method?

Ans.: $1630.

**3-28** Regular payments of $1400 are to be made annually, starting at the end of the first year. These amounts will be invested at 6 percent compounded continuously. How many years will be needed for the payments plus interest to accumulate to $24,000?

Ans.: 12 years.

**3-29** An investment of $300,000 yields an annual profit of $86,000 that is spread uniformly over the year and is reinvested immediately (thus continuously compounded). The life is 6 years, and there is no salvage value. What is the rate of return on the investment?
**Ans.:** 20%.

# ADDITIONAL READINGS

Barish, N. N.: *Economic Analysis for Engineering and Managerial Decision Making*, McGraw-Hill, New York, 1962.

DeGarmo, E. P.: *Engineering Economy*, Macmillan, New York, 1967.

Grant, E. L., and W. G. Ireson: *Principles of Engineering Economy*, 4th ed., Ronald Press, New York, 1960.

Smith, G. W.: *Engineering Economy*, Iowa State University Press, Ames, 1968.

Taylor, G. A.: *Managerial and Engineering Economy*, Van Nostrand, Princeton, N.J., 1964.

# FOUR

## EQUATION FITTING

## 4-1 MATHEMATICAL MODELING

This chapter and the next present procedures for developing equations that represent the performance characteristics of equipment, the behavior of processes, and thermodynamic properties of substances. Engineers may have a variety of reasons for wanting to develop equations, but the crucial ones in the design of thermal systems are (1) to facilitate the process of system simulation and (2) to develop a mathematical statement for optimization. Most large, realistic simulation and optimization problems must be executed on the computer, and it is usually more expedient to operate with equations than with tabular data. An emerging need for expressing equations is in *equipment selection*; some designers are automating equipment selection, storing performance data in the computer, and then automatically retrieving them when a component is being selected.

Equation development will be divided into two different categories; this chapter treats equation fitting and Chap. 5 concentrates on modeling thermal equipment. The distinction between the two is that this chapter approaches the development of equations as purely a number-processing operation, while Chap. 5 uses some physical laws to help equation development. Both approaches are appropriate. In modeling a reciprocating compressor, for example, obviously there are physical explanations for the performance, but by the time the complicated flow processes, compression, reexpansion, and valve mechanics are incorporated, the model is so complex that it is simpler to use experimental or catalog data and treat the problem as a number-processing exercise. On the other hand, heat exchangers follow certain laws that suggest a form for the equation, and this insight can be used to advantage, as shown in Chap. 5.

Where do the data come from on which equations are based? Usually the data used by a designer come from tables or graphs. Experimental data from the laboratory might provide the basis, and the techniques in this and the next chapter are applicable to processing laboratory data. But system designers are usually one step removed from the laboratory and are selecting commercially available components for which the manufacturer has provided performance data. In a few rare instances manufacturers may reserve several lines on a page of tabular data to provide the equation that represents the table. If and when that practice becomes widespread, the system designer's task will be made easier. That stage, however, has not yet been reached.

Much of this chapter presents systematic techniques for determining the constants and coefficients in equations, a process of following rules. The other facet of equation fitting is that of proposing the form of the equation, and this operation is an art. Some suggestions will be offered for the execution of this art. Methods will be presented for determining equations that fit a limited number of data points perfectly. Also explained is the method of least squares, which provides an equation of best fit to a large number of points.

## 4-2 MATRICES

All the operations in this chapter can be performed without using matrix terminology, but the use of matrices provides several conveniences and insights. In particular, the application of matrix terminology is applicable to the solution of sets of simultaneous equations.

A matrix is a rectangular array of numbers, for example,

$$\begin{bmatrix} 5 & -2 & 0 \\ 3 & 1 & -1 \\ 0 & 1 & 1 \end{bmatrix} \quad \begin{bmatrix} 2 & 3 \\ -1 & 0 \\ 2 & 2 \end{bmatrix} \quad \begin{bmatrix} 7 & 3 \\ 1 & 2 \end{bmatrix} \quad \begin{bmatrix} a_{11} & a_{12} & \cdots & a_{1n} \\ a_{21} & a_{22} & \cdots & a_{2n} \\ \cdots\cdots\cdots\cdots\cdots\cdots \\ a_{m1} & a_{m2} & \cdots & a_{mn} \end{bmatrix}$$

The numbers that make up the array are called *elements*. The orders of these matrices, from left to right, are $3 \times 3$, $3 \times 2$, $2 \times 2$, and $m \times n$.

A transpose of a matrix $[A]$, designated $[A]^T$, is formed by interchanging rows and columns. Thus, if

$$[A] = \begin{bmatrix} 3 & -1 \\ 2 & 0 \\ 4 & -2 \end{bmatrix} \quad \text{then} \quad [A]^T = \begin{bmatrix} 3 & 2 & 4 \\ -1 & 0 & -2 \end{bmatrix}$$

To multiply two matrices, multiply elements of the first row of the left matrix by the corresponding elements of the first column of the right matrix; then sum the products to give the element of the first row and first column of the product matrix. For example, the multiplication of the two matrices

$$\begin{bmatrix} 1 & -1 & 0 \\ 2 & 0 & 1 \end{bmatrix} \quad \text{and} \quad \begin{bmatrix} -2 & 1 \\ 3 & 0 \\ 1 & 4 \end{bmatrix}$$

gives

$$\begin{bmatrix} (1)(-2)+(-1)(3)+(0)(1) & (1)(1)+(-1)(0)+(0)(4) \\ (2)(-2)+(0)(3)+(1)(1) & (2)(1)+(0)(0)+(1)(4) \end{bmatrix} = \begin{bmatrix} -5 & 1 \\ -3 & 6 \end{bmatrix}$$

The convention for the multiplication of two matrices offers a slightly shorter form for writing a system of simultaneous linear equations. The three equations

$$\begin{aligned} 2x_1 - x_2 + 3x_3 &= 6 \\ x_1 + 3x_2 \quad\quad &= 1 \\ 4x_1 - 2x_2 + x_3 &= 0 \end{aligned}$$

can be written in matrix form as

$$\begin{bmatrix} 2 & -1 & 3 \\ 1 & 3 & 0 \\ 4 & -2 & 1 \end{bmatrix} \begin{bmatrix} x_1 \\ x_2 \\ x_3 \end{bmatrix} = \begin{bmatrix} 6 \\ 1 \\ 0 \end{bmatrix}$$

The determinant is a scalar (which is simply a number) and is written between vertical lines. For a $1 \times 1$ matrix it is the element itself; thus

$$|a_{11}| = a_{11}$$

A technique for evaluating the determinant of a $2 \times 2$ matrix is to sum the products of diagonal elements, assigning a positive sign to the diagonal moving downward to the right and a negative sign to the product moving upward to the right:

$$\begin{vmatrix} a_{11} & a_{12} \\ a_{21} & a_{22} \end{vmatrix} = + \searrow - \nearrow = a_{11}a_{22} - a_{21}a_{12}$$

An extension of this method applies to computing the determinant of a $3 \times 3$ matrix:

$$\begin{vmatrix} a_{11} & a_{12} & a_{13} \\ a_{21} & a_{22} & a_{23} \\ a_{31} & a_{32} & a_{33} \end{vmatrix} = + \searrow + \searrow + \searrow - \nearrow - \nearrow - \nearrow$$

$$= a_{11}a_{22}a_{33} + a_{12}a_{23}a_{31} + a_{13}a_{21}a_{32}$$

$$- a_{31}a_{22}a_{13} - a_{32}a_{23}a_{11} - a_{33}a_{21}a_{12}$$

Evaluation of determinants $4 \times 4$ and larger requires a more general procedure, which applies also to $2 \times 2$ and $3 \times 3$ matrices. This procedure is *row expansion* or *column expansion*. The determinant of a $3 \times 3$ matrix found by expanding about the first column is

$$\det = a_{11}A_{11} + a_{21}A_{21} + a_{31}A_{31}$$

where $A_{ij}$ is the *cofactor* of the element $a_{ij}$. The cofactor is found as follows:

$$A_{ij} = [(-1)^{i+j}] \begin{vmatrix} \text{submatrix formed} \\ \text{by striking out} \\ i\text{th row and } j\text{th} \\ \text{column of } [A] \end{vmatrix}$$

For example, the cofactor of $a_{21}$, which is $A_{21}$, is

$$A_{21} = [(-1)^{2+1}] \begin{vmatrix} a_{11} & a_{12} & a_{13} \\ a_{21} - a_{22} - a_{23} \\ a_{31} & a_{32} & a_{33} \end{vmatrix} = (-1)^3 \begin{vmatrix} a_{12} & a_{13} \\ a_{32} & a_{33} \end{vmatrix}$$

$$A_{21} = -(a_{12}a_{33} - a_{32}a_{13})$$

**Example 4-1**  Evaluate

$$\begin{vmatrix} 1 & 2 & -1 & 0 \\ 0 & 1 & 2 & 0 \\ 3 & -1 & 1 & 2 \\ 4 & 2 & 1 & 5 \end{vmatrix}$$

SOLUTION  Two elements of the second row are zero, so that row would be a convenient one about which to expand.

$$\det = a_{21}A_{21} + a_{22}A_{22} + a_{23}A_{23} + a_{24}A_{24}$$

$$= (0)A_{21} + (1)(-1)^{2+2} \begin{vmatrix} 1 & -1 & 0 \\ 3 & 1 & 2 \\ 4 & 1 & 5 \end{vmatrix}$$

$$+ (2)(-1)^{2+3} \begin{vmatrix} 1 & 2 & 0 \\ 3 & -1 & 2 \\ 4 & 2 & 5 \end{vmatrix} + (0)A_{24}$$

$$= 0 + 10 + 46 + 0 = 56$$

# 4-3 SOLUTION OF SIMULTANEOUS EQUATIONS

There are many ways of solving sets of simultaneous equations, two of which will be described in this section, Cramer's rule and gaussian elimination. For a set of linear simultaneous equations

$$\begin{aligned} a_{11}x_1 + a_{12}x_2 + a_{13}x_3 &= b_1 \\ a_{21}x_1 + a_{22}x_2 + a_{23}x_3 &= b_2 \\ a_{31}x_1 + a_{32}x_2 + a_{33}x_3 &= b_3 \end{aligned} \qquad (4\text{-}1)$$

which can be written in matrix form

$$[A] [X] = \begin{bmatrix} a_{11} & a_{12} & a_{13} \\ a_{21} & a_{22} & a_{23} \\ a_{31} & a_{32} & a_{33} \end{bmatrix} \begin{bmatrix} x_1 \\ x_2 \\ x_3 \end{bmatrix} = \begin{bmatrix} b_1 \\ b_2 \\ b_3 \end{bmatrix} = [B] \tag{4-2}$$

Cramer's rule states that

$$x_i = \frac{|\ [A]\ \text{matrix with } [B]\ \text{matrix substituted in } i\text{th column}\ |}{|A|} \tag{4-3}$$

**Example 4-2** Using Cramer's rule, solve for $x_2$ in this set of simultaneous linear equations:

$$\begin{bmatrix} 2 & 1 & -1 \\ 1 & -2 & 2 \\ -1 & 0 & 3 \end{bmatrix} \begin{bmatrix} x_1 \\ x_2 \\ x_3 \end{bmatrix} = \begin{bmatrix} 3 \\ 9 \\ 0 \end{bmatrix}$$

SOLUTION

$$x_2 = \frac{\begin{vmatrix} 2 & 3 & -1 \\ 1 & 9 & 2 \\ -1 & 0 & 3 \end{vmatrix}}{\begin{vmatrix} 2 & 1 & -1 \\ 1 & -2 & 2 \\ -1 & 0 & 3 \end{vmatrix}} = \frac{30}{-15} = -2$$

Equation (4-3) suggests that none of the $x$'s can be determined if $|A|$ is zero. The equations are dependent in this case, and there is no unique solution to the set.

Another method of solving simultaneous linear equations is gaussian elimination, which will be illustrated by solving

$$x_1 - 4x_2 + 3x_3 = -7 \tag{4-4}$$

$$3x_1 + x_2 - 2x_3 = 14 \tag{4-5}$$

$$2x_1 + x_2 + x_3 = 5 \tag{4-6}$$

The two major steps in gaussian elimination are conversion of the coefficient matrix into a triangular matrix and solution for $x_n$ to $x_1$ by back substitution.

In the example set of equations, the first part of step 1 is to eliminate the coefficients of $x_1$ in Eq. (4-5) by multiplying Eq. (4-4) by a suitable constant and adding the product to Eq. (4-5). Specifically, multiply Eq. (4-4) by $-3$ and add to Eq. (4-5). Similarly, multiply Eq. (4-4) by $-2$ and add to Eq. (4-6):

$$x_1 - 4x_2 + 3x_3 = -7 \tag{4-7}$$

$$13x_2 - 11x_3 = 35 \tag{4-8}$$

$$9x_2 - 5x_3 = 19 \tag{4-9}$$

The last part of step 1 is to multiply Eq. (4-8) by $-\frac{9}{13}$ and add to Eq. (4-9), which completes the triangularization

$$x_1 - 4x_2 + 3x_3 = -7 \qquad (4\text{-}10)$$

$$13x_2 - 11x_3 = 35 \qquad (4\text{-}11)$$

$$\tfrac{34}{13}x_3 = -\tfrac{68}{13} \qquad (4\text{-}12)$$

In step 2 the value of $x_3$ can be determined directly from Eq. (4-12) as $x_3 = -2$. Substituting the value of $x_3$ into Eq. (4-11) and solving gives $x_2 = 1$. Finally, substitute the values of $x_2$ and $x_3$ into Eq. (4-10) to find that $x_1 = 3$.

If a different set of equations were being solved, and in the equation corresponding to Eq. (4-8) if the coefficient of $x_2$ had been zero instead of 13, it would have been necessary to exchange the positions of Eqs. (4-8) and (4-9). If both the $x_2$ coefficients in Eqs. (4-8) and (4-9) had been zero, this would indicate that the set of equations is dependent.

Most computer departments have in their library a routine for solving a set of simultaneous linear equations which can be called as needed. It may be convenient to write one's own subprogram using a method like gaussian elimination.[1] It will be useful for future work in this text to have access to an equation-solving routine on a digital computer.

## 4-4 POLYNOMIAL REPRESENTATIONS

Probably the most obvious and most useful form of equation representation is a polynomial. If $y$ is to be represented as a function of $x$, the polynomial form is

$$y = a_0 + a_1 x + a_2 x^2 + \cdots + a_n x^n \qquad (4\text{-}13)$$

where $a_0$ to $a_n$ are constants. The degree of the equation is the highest exponent of $x$, which in Eq. (4-13) is $n$.

Equation (4-13) is an expression giving the function of one variable in terms of another. In other common situations one variable is a function of two or more variables, e.g., in an axial-flow compressor

Flow rate = $f$ (inlet pressure, inlet temperature, compressor speed, outlet pressure)

This form of equation will be presented in Sec. 4-8.

When the number of data points available is precisely the same as the degree of the equation plus 1, $n + 1$, a polynomial can be devised that exactly expresses those data points. When the number of available data points exceeds $n + 1$, it may be advisable to seek a polynomial that gives the "best fit" to the data points (see Sec. 4-10).

The first and simplest case to be considered is where one variable is a function of another variable and the number of data points equals $n + 1$.

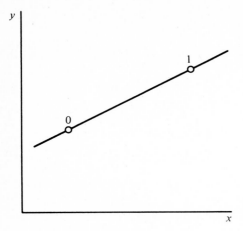

**Figure 4-1** Two points describing a linear equation.

## 4-5 POLYNOMIAL, ONE VARIABLE A FUNCTION OF ANOTHER VARIABLE AND $n + 1$ DATA POINTS

Two available data points are adequate to describe a first-degree, or linear, equation (Fig. 4-1). The form of this first-degree equation is

$$y = a_0 + a_1 x \tag{4-14}$$

The $xy$ pairs for the two known points $(x_0, y_0)$ and $(x_1, y_1)$ can be substituted into Eq. (4-14), providing two linear equations with two unknowns, $a_0$ and $a_1$

$$y_0 = a_0 + a_1 x_0$$

$$y_1 = a_0 + a_1 x_1$$

For a second-degree, or quadratic, equation, three data points are needed; for example, points 0, 1, and 2 in Fig. 4-2. The $xy$ pairs for the three known

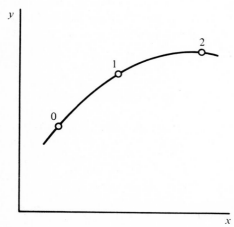

**Figure 4-2** Three points describing a quadratic equation.

points can be substituted into the general form for the quadratic equation

$$y = a_0 + a_1 x + a_2 x^2 \qquad (4\text{-}15)$$

which gives three equations

$$\begin{bmatrix} 1 & x_0 & x_0^2 \\ 1 & x_1 & x_1^2 \\ 1 & x_2 & x_2^2 \end{bmatrix} \begin{bmatrix} a_0 \\ a_1 \\ a_2 \end{bmatrix} = \begin{bmatrix} y_0 \\ y_1 \\ y_2 \end{bmatrix}$$

The solution of these three linear simultaneous equations provides the values of $a_0, a_1$, and $a_2$.

The coefficients of the high-degree terms in a polynomial may be quite small, particularly if the independent variable is large. For example, if the enthalpy of saturated water vapor $h$ is a function of temperature $t$ in the equation

$$h = a_0 + a_1 t + \cdots + a_5 t^5 + a_6 t^6$$

where the range of $t$ extends into hundreds of degrees, the value of $a_5$ and $a_6$ may be so small that precision problems result. Sometimes this difficulty can be surmounted by defining a new independent variable, for example, $t/100$.

$$h = a_0 + a_1 \frac{t}{100} + \cdots + a_5 \left(\frac{t}{100}\right)^5 + a_6 \left(\frac{t}{100}\right)^6$$

## 4-6 SIMPLIFICATIONS WHEN THE INDEPENDENT VARIABLE IS UNIFORMLY SPACED

Sometimes a polynomial is used to represent a function, say $y = f(x)$, where the values of $y$ are known at equally spaced values of $x$. This situation exists, for instance, when the data points are read off a graph and the points can be chosen at equal intervals of $x$. The solution of simultaneous equations to determine the coefficients in the polynomial can be performed symbolically in advance,[2] and the execution of the calculations requires a relatively small effort thereafter.

Suppose that the curve in Fig. 4-3 is to be reproduced by a fourth-degree polynomial. The $n + 1$ data points (five in this case) establish a polynomial of degree $n$ (four in this case). The spacing of the points is $x_1 - x_0 = x_2 - x_1 = x_3 - x_2 = x_4 - x_3$. The range of $x$, $x_4 - x_0$, is designated $R$, and the symbols are $\Delta y_1 = y_1 - y_0$, $\Delta y_2 = y_2 - y_0$, etc.

Instead of the polynomial form of Eq. (4-13), an alternate form is used

$$y - y_0 = a_1 \left[ \frac{n}{R} (x - x_0) \right] + a_2 \left[ \frac{n}{R} (x - x_0) \right]^2 + a_3 \left[ \frac{n}{R} (x - x_0) \right]^3$$

$$+ a_4 \left[ \frac{n}{R} (x - x_0) \right]^4 \qquad (4\text{-}16)$$

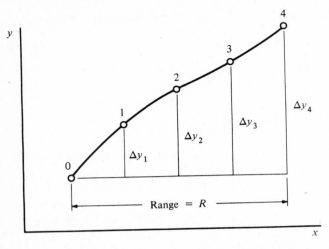

**Figure 4-3** Polynomial representation when points are equally spaced along the $x$ axis.

To find $a_1$ to $a_4$, first substitute the $(x_1, y_1)$ pair into Eq. (4-16)

$$\Delta y_1 = a_1 \frac{4(x_1 - x_0)}{R} + a_2 \left[\frac{4(x_1 - x_0)}{R}\right]^2 + a_3 \left[\frac{4(x_1 - x_0)}{R}\right]^3 + a_4 \left[\frac{4(x_1 - x_0)}{R}\right]^4$$

(4-17)

Because of the uniform spacing of the points along the $x$ axis, $n(x_1 - x_0)/R = 1$, and so Eq. (4-17) can be rewritten as

$$\Delta y_1 = a_1 + a_2 + a_3 + a_4$$

(4-18)

**Table 4-1  Constants in Eq. (4-16)**

| Equation | $a_4$ | $a_3$ | $a_2$ | $a_1$ |
|---|---|---|---|---|
| Fourth degree | $\frac{1}{24}(\Delta y_4 - 4\Delta y_3 + 6\Delta y_2 - 4\Delta y_1)$ | $\frac{\Delta y_3}{6} - \frac{\Delta y_2}{2} + \frac{\Delta y_1}{2} - 6a_4$ | $\frac{\Delta y_2}{2} - \Delta y_1 - 3a_3 - 7a_4$ | $\Delta y_1 - a_2 - a_3 - a_4$ |
| Cubic | | $\frac{1}{6}(3\Delta y_1 + \Delta y_3 - 3\Delta y_2)$ | $\frac{1}{2}(\Delta y_2 - 2\Delta y_1) - 3a_3$ | $\Delta y_1 - a_2 - a_3$ |
| Quadratic | | | $\frac{1}{2}(\Delta y_2 - 2\Delta y_1)$ | $\Delta y_1 - a_2$ |
| Linear | | | | $\Delta y_1$ |

Using the $(x_2, y_2)$ pair and the fact that $n(x_2 - x_0)/R = 2$ gives

$$\Delta y_2 = 2a_1 + 4a_2 + 8a_3 + 16a_4 \tag{4-19}$$

Similarly, for $(x_3, y_3)$ and $(x_4, y_4)$

$$\Delta y_3 = 3a_1 + 9a_2 + 27a_3 + 81a_4 \tag{4-20}$$

$$\Delta y_4 = 4a_1 + 16a_2 + 64a_3 + 256a_4 \tag{4-21}$$

The expressions for $a_1$ to $a_4$ found by solving Eqs. (4-18) to (4-21) simultaneously are shown in Table 4-1, along with the constants for the cubic, quadratic, and linear equations.

## 4-7 LAGRANGE INTERPOLATION

Another form of polynomial results when using Lagrange interpolation. This method is applicable, unlike the method described in Sec. 4-6, to arbitrary spacing along the $x$ axis. It has the advantage of not requiring the simultaneous solution of equations but is cumbersome to write out. This disadvantage is not applicable if the calculation is performed on a digital computer, in which case the programming is quite compact.

With a quadratic equation as an example, the usual form for a function of one variable is

$$y = a_0 + a_1 x + a_2 x^2 \tag{4-22}$$

For Lagrange interpolation, a revised form is used

$$y = c_1(x - x_2)(x - x_3) + c_2(x - x_1)(x - x_3) + c_3(x - x_1)(x - x_2) \tag{4-23}$$

The three available data points are $(x_1, y_1)$, $(x_2, y_2)$, and $(x_3, y_3)$. Equation (4-23) could be multiplied out and terms collected to show the correspondence to the form in Eq. (4-22).

By setting $x = x_1$, $x_2$, and $x_3$ in turn in Eq. (4-23) the constants can be found quite simply

$$c_1 = \frac{y_1}{(x_1 - x_2)(x_1 - x_3)}$$

$$c_2 = \frac{y_2}{(x_2 - x_1)(x_2 - x_3)}$$

$$c_3 = \frac{y_3}{(x_3 - x_1)(x_3 - x_2)}$$

The general form of the equation for finding the value of $y$ for a given $x$ when $n$ data points are known is

$$y = \sum_{i=1}^{n} y_i \prod_{j=1}^{n} \frac{(x - x_j) \text{ omitting } (x - x_i)}{(x_i - x_j) \text{ omitting } (x_i - x_i)} \qquad (4\text{-}24)$$

where the pi, or product sign, indicates multiplication.

The equation represented by Eq. (4-24) is a polynomial of degree $n - 1$.

## 4-8 FUNCTION OF TWO VARIABLES

A performance variable of a component is often a function of two other variables,[3] not just one. For example, the pressure rise developed by the centrifugal pump shown in Fig. 4-4 is a function of both the speed $S$ and the flow rate $Q$.

If a polynomial expression for the pressure rise $\Delta p$ is sought in terms of a second-degree equation in $S$ and $Q$, separate equations can be written for each of the three curves in Fig. 4-4. Three points on the 30 r/s curve would provide the constants in the equation

$$\Delta p_1 = a_1 + b_1 Q + c_1 Q^2 \qquad (4\text{-}25)$$

Similar equations for the curves for the 24 and 16 r/s speeds are

$$\Delta p_2 = a_2 + b_2 Q + c_2 Q^2 \qquad (4\text{-}26)$$

$$\Delta p_3 = a_3 + b_3 Q + c_3 Q^2 \qquad (4\text{-}27)$$

Next the $a$ constants can be expressed as a second-degree equation in terms of $S$, using the three data points $(a_1, 30)$, $(a_2, 24)$, and $(a_3, 16)$. Such an equation

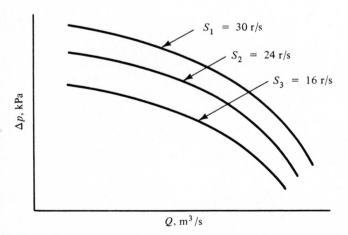

Figure 4-4 Performance of a centrifugal pump.

would have the form

$$a = A_0 + A_1 S + A_2 S^2 \qquad (4\text{-}28)$$

Similarly for $b$ and $c$

$$b = B_0 + B_1 S + B_2 S^2 \qquad (4\text{-}29)$$

$$c = C_0 + C_1 S + C_2 S^2 \qquad (4\text{-}30)$$

Finally, the constants of Eqs. (4-28) to (4-30) are put into the general equation

$$\Delta p = A_0 + A_1 S + A_2 S^2 + (B_0 + B_1 S + B_2 S^2) Q + (C_0 + C_1 S + C_2 S^2) Q^2 \qquad (4\text{-}31)$$

The $A$, $B$, and $C$ constants can be computed if nine data points from Fig. 4-4 are available.

**Example 4-3** Manufacturers of cooling towers often present catalog data showing the outlet-water temperature as a function of the wet-bulb temperature of the ambient air and the range. The range is the difference between the inlet and outlet temperatures of the water. In Table 4-2, for example, when the wet-bulb temperature is $20°C$ and the range is $10°C$, the temperature of leaving water is $25.9°C$, and so the temperature of the entering water is $25.9 + 10 = 35.9°C$. Express the outlet-water temperature $t$ in Table 4-2 as a function of the wet-bulb temperature (WBT) and the range $R$.

SOLUTION A second-degree polynomial equation in both independent variables will be chosen as the form of the equation, and three different methods for developing the equation will be illustrated.

*Method 1* The three pairs of points for WBT = $20°C$, (10, 25.9), (16, 27.0), and (22, 28.4), can be represented by a parabola

$$t = 24.733 + 0.075006R + 0.004146R^2$$

For WBT = $23°C$

$$t = 26.667 + 0.041659R + 0.0041469R^2$$

and for WBT = $26°C$

$$t = 28.733 + 0.024999R + 0.0041467R^2$$

**Table 4-2  Outlet-water temperature, °C, of cooling tower in Example 4-3**

| Range, °C | Wet-bulb temperature, °C | | |
|---|---|---|---|
| | 20 | 23 | 26 |
| 10 | 25.9 | 27.5 | 29.4 |
| 16 | 27.0 | 28.4 | 30.2 |
| 22 | 28.4 | 29.6 | 31.3 |

Next, the constant terms 24.733, 26.667, and 28.733 can be expressed by a second-degree equation of WBT,

$$15.247 + 0.32637\text{WBT} + 0.007380\text{WBT}^2$$

The coefficients of $R$ and $R^2$ can also be expressed by equations in terms of the WBT, which then provide the complete equation

$$t = (15.247 + 0.32637\text{WBT} + 0.007380\text{WBT}^2)$$
$$+ (0.72375 - 0.050978\text{WBT} + 0.000927\text{WBT}^2)R$$
$$+ (0.004147 + 0\text{WBT} + 0\text{WBT}^2)R^2 \qquad (4\text{-}32)$$

*Method 2* An alternate polynomial form using second-degree expressions for $R$ and WBT is

$$t = c_1 + c_2\text{WBT} + c_3\text{WBT}^2 + c_4R + c_5R^2 + c_6(R)(\text{WBT})$$
$$+ c_7(\text{WBT})^2(R) + c_8(\text{WBT})(R)^2 + c_9(\text{WBT})^2(R)^2 \qquad (4\text{-}33)$$

The nine sets of $t$-$R$-WBT combinations expressed in Table 4-2 can be substituted into Eq. (4-33) to develop nine simultaneous equations, which can be solved for the unknowns $c_1$ to $c_9$. The $c$ values thus obtained are

$$
\begin{array}{lll}
c_1 = 15.247 & c_2 = 0.32631 & c_3 = 0.0073991 \\
c_4 = 0.723753 & c_5 = 0.0041474 & c_6 = -0.0509782 \\
c_7 = 0.00092704 & c_8 = 0.0 & c_9 = 0.0
\end{array}
$$

It is possible to multiply and collect the terms in Eq. (4-32) to develop the equation of the form of Eq. (4-33).

*Method 3* Section 4-7 described a polynomial representation of a dependent variable as a function of one independent variable by use of Lagrange interpolation. Lagrange interpolation can be extended to a function of two independent variables. For example, if $z = f(x, y)$, the form can be chosen

$$z = c_{11}(x - x_2)(x - x_3)(y - y_2)(y - y_3)$$
$$+ c_{12}(x - x_2)(x - x_3)(y - y_1)(y - y_3)$$
$$+ c_{13}(x - x_2)(x - x_3)(y - y_1)(y - y_2)$$
$$+ \cdots + c_{33}(x - x_1)(x - x_2)(y - y_1)(y - y_2) \qquad (4\text{-}34)$$

To represent the data in Table 4-2, $z$ could refer to the outlet-water temperature, $x$ the WBT, and $y$ the range. In Eq. (4-34) $x_1 = 20$, $x_2 = 23$, and $x_3 = 26$, while $y_1 = 10$, $y_2 = 16$, and $y_3 = 22$.

To determine the magnitude of $c_{12}$, for example, values applicable when $x = x_1$ and $y = y_2$ can be substituted into Eq. (4-34).

$$c_{12} = \frac{27.0}{(20 - 23)(20 - 26)(16 - 10)(16 - 22)} = -0.04167$$

## 4-9  EXPONENTIAL FORMS

The dependence of one variable on a second variable raised to an exponent is a physical relation occurring frequently in engineering practice. The graphical method of determining the constants $b$ and $m$ in the equation

$$y = bx^m \tag{4-35}$$

is a simple example of mathematical modeling of an exponential form. On a graph of the known values of $x$ and $y$ on a log-log plot (Fig. 4-5) the slope of the straight line through the points equals $m$, and the intercept at $x = 1$ defines $b$.

The simple exponential form of Eq. (4-35) can be extended to include a constant

$$y = b + ax^m \tag{4-36}$$

The equation permits representations of curves similar to those shown in Fig. 4-6. The curve shown in Fig. 4-6b is especially common in engineering practice. The function $y$ approaches some value $b$ asymptotically as $x$ increases.

One possible graphical method of determining $a$, $b$, and $m$ in Eq. (4-36) when pairs of $xy$ values are known is as follows:

1. Estimate the value of $b$.
2. Use the steep portion of the curve to evaluate $m$ by a log-log plot of $y - b$ vs. $x$ in a manner similar to that shown in Fig. 4-5.
3. With the value of $m$ from step 2, plot a graph of $y$ vs. $x^m$. The resulting curve should be a straight line with a slope of $a$ and an intercept that indicates a more correct value of $b$. Iterate starting at step 2 if desired.

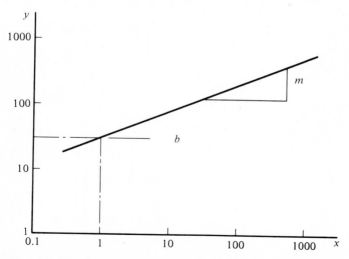

Figure 4-5  Graphical determination of the constant $b$ and exponent $m$.

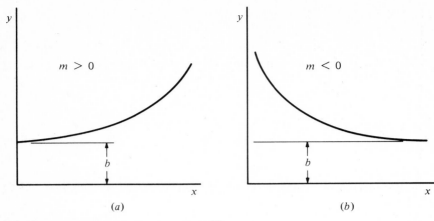

**Figure 4-6** Curve of the equation $y = b + ax^m$.

## 4-10 BEST FIT: METHOD OF LEAST SQUARES

This chapter has concentrated so far on finding equations that give a perfect fit to a limited number of points. If $m$ coefficients are to be determined in an equation, $m$ data points are required. If more than $m$ points are available, it is possible to determine the $m$ coefficients that result in the best fit of the equation to the data. One definition of a best fit is the one where the sum of the absolute values of the deviations from the data points is a minimum. In another type of best fit slightly different from the one just mentioned the sum of the squares of the deviation is a minimum. The procedure in establishing the coefficients in such an equation is called the *method of least squares*.

Some people proudly announce their use of the method of least squares in order to emphasize the care they have lavished on their data analysis. Misuses of the method, as illustrated in Fig. 4-7$a$ and $b$, are not uncommon. In Fig. 4-7$a$, while a straight line can be found that results in the least-squares deviation, the correlation between the $x$ and $y$ variables seems questionable and perhaps no such device can improve the correlation. The scatter may be due to the omission of some significant variable(s). In Fig. 4-7$b$ it would have been preferable to eyeball in the curve, rather than to fit a straight line to the data by the least-squares method. The error was not in using least squares but in applying a curve of too low a degree.

The procedure for using the least-squares method for first- and second-degree polynomials will be explained here. Consider first the linear equation of the form

$$y = a + bx \tag{4-37}$$

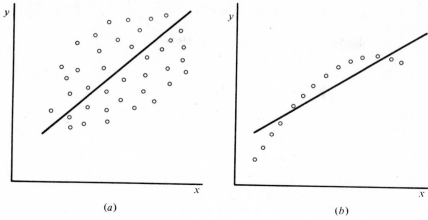

Figure 4-7 Misuses of the method of least squares.

where $m$ pairs of data points are available: $(x_1, y_1), (x_2, y_2), \ldots, (x_m, y_m)$. The deviation of the data point from that calculated from the equation is $a + bx_i - y_i$. We wish to choose an $a$ and a $b$ such that the summation

$$\sum_{i=1}^{m} (a + bx_i - y_i)^2 \longrightarrow \text{minimum} \qquad (4\text{-}38)$$

The minimum occurs when the partial derivatives of Eq. (4-38) with respect to $a$ and $b$ equal zero.

$$\frac{\partial \sum_{i=1}^{m} (a + bx_i - y_i)^2}{\partial a} = \Sigma\, 2(a + bx_i - y_i) = 0$$

and

$$\frac{\partial \sum_{i=1}^{m} (a + bx_i - y_i)^2}{\partial b} = \Sigma\, 2(a + bx_i - y_i)\, x_i = 0$$

Dividing by 2 and separating the above two equations into individual terms gives

$$ma + b\, \Sigma\, x_i = \Sigma\, y_i \qquad (4\text{-}39)$$

$$a\, \Sigma\, x_i + b\, \Sigma\, x_i^2 = \Sigma\, x_i y_i \qquad (4\text{-}40)$$

**Example 4-4** Determine $a_0$ and $a_1$ in the equation $y = a_0 + a_1 x$ to provide a best fit in the sense of least-squares deviation to the data points $(1, 4.9)$, $(3, 11.2)$, $(4, 13.7)$, and $(6, 20.1)$.

SOLUTION  The summations to substitute into Eqs. (4-39) and (4-40) are

| | $x_i$ | $y_i$ | $x_i^2$ | $x_i y_i$ |
|---|---|---|---|---|
| | 1 | 4.9 | 1 | 4.9 |
| | 3 | 11.2 | 9 | 33.6 |
| | 4 | 13.7 | 16 | 54.8 |
| | 6 | 20.1 | 36 | 120.6 |
| $\Sigma$ | 14 | 49.9 | 62 | 213.9 |

and  $m = 4$.

The simultaneous equations to be solved are

$$4a_0 + 14a_1 = 49.9$$
$$14a_0 + 62a_1 = 213.9$$

yielding $a_0 = 1.908$ and $a_1 = 3.019$. Thus

$$y = 1.908 + 3.019x$$

A similar procedure can be followed when fitting a parabola of the form

$$y = a + bx + cx^2 \tag{4-41}$$

to $m$ data points. The summation to be minimized is

$$\sum_{i=1}^{m} (a + bx_i + cx_i^2 - y_i)^2 \longrightarrow \text{minimum}$$

Differentiating partially with respect to $a$, $b$, and $c$, in turn, results in three linear simultaneous equations expressed in matrix form

$$\begin{bmatrix} m & \Sigma x_i & \Sigma x_i^2 \\ \Sigma x_i & \Sigma x_i^2 & \Sigma x_i^3 \\ \Sigma x_i^2 & \Sigma x_i^3 & \Sigma x_i^4 \end{bmatrix} \begin{bmatrix} a \\ b \\ c \end{bmatrix} = \begin{bmatrix} \Sigma y_i \\ \Sigma x_i y_i \\ \Sigma x_i^2 y_i \end{bmatrix} \tag{4-42}$$

A comparison of the matrix equation (4-42) with Eqs. (4-39) and (4-40) shows a pattern evolving which by analogy permits developing the equations for higher degree polynomials without even differentiating the summation of the squared deviation.

## 4-11 METHOD OF LEAST SQUARES APPLIED TO NONPOLYNOMIAL FORMS

The explanation of the method of least squares was applied to polynomial forms in Sec. 4-10, but it should not be suggested that the method is limited to those forms. The method is applicable to any form which contains constant coefficients. For example, if the form of the equation is

$$y = a \sin 2x + b \ln x^2$$

the summation comparable to Eq. (4-38) is

$$\sum_{i=1}^{m} (y_i - a \sin 2x_i - b \ln x_i^2)^2 \qquad (4\text{-}43)$$

Partial differentiation with respect to $a$ and $b$ yields

$$a \, \Sigma \, (\sin 2x_i)^2 + b \, \Sigma \, (\sin 2x_i)(\ln x_i^2) = \Sigma \, y_i \sin 2x_i$$

$$a \, \Sigma \, (\sin 2x_i)(\ln x_i^2) + b \, \Sigma \, (\ln x_i^2)^2 = \Sigma \, y_i \ln x_i^2$$

which can be solved for $a$ and $b$.

A crucial characteristic of the equation form that makes it tractable to the method of least squares is that the equation have constant coefficients. In an equation of the form

$$y = \sin 2ax + bx^c$$

the terms $a$ and $c$ do not appear as coefficients, and this equation cannot be handled in a straightforward manner by least squares.

## 4-12 THE ART OF EQUATION FITTING

While there are methodical procedures for fitting equations to data, the process is also an art. The art or intuition is particularly needed in deciding upon the form of the equation, namely, the choice of independent variables to be included and the form in which these variables should appear. There are no fixed rules for knowing what variables to include or what their form should be in the equation,

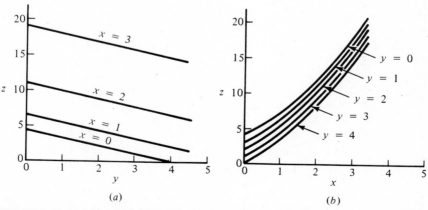

**Figure 4-8** Cross plots to aid in developing the form of the equation.

but making at least a rough plot of the data will often provide some insight. If the dependent variable is a function of two independent variables, as in $z = f(x, y)$, two plots might be made, as illustrated in Fig. 4-8.

The insight provided by Fig. 4-8$a$ is that $z$ bears a linear relation to $y$, and the fact that the straight lines are parallel shows no influence of $x$ on the slope. Figure 4-8$b$ suggests at least a second-degree representation of $z$ as a function of $x$. A reasonable form to propose, then, is

$$z = a_0 + a_1 x + a_2 y + a_3 x^2$$

Several frequently used forms merit further discussion.

## Polynomials

If there is a lack of special indicators that other forms are more applicable, a polynomial would probably be explored. When the curve has a reverse curvature (inflection point), as shown in Fig. 4-9, at least a third-degree polynomial must be chosen. Extrapolation of a polynomial beyond the borders of the data used to develop the equation often results in serious error.

## Polynomials with Negative Exponents

When curves approach a constant value at large magnitudes of the independent variable, polynomials with negative exponents

$$y = a_0 + a_1 x^{-1} + a_2 x^{-2}$$

may provide a good representation; see Fig. 4-10.

## Exponential Equations

Section 4-9 has described several examples of exponential forms. The shape of the curve in Fig. 4-10 might also include a $c^{-x}$ term. Plots on log-log paper would

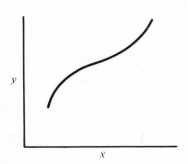

**Figure 4-9** At least a third-degree polynomial needed.

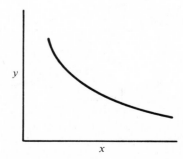

**Figure 4-10** Negative exponents of polynomials for a curve that flattens out.

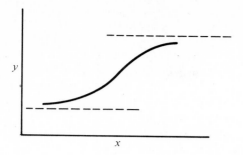

**Figure 4-11** Gompertz, or $S$ curve.

be a routine procedure, although a simple plot of $\log y$ vs. $\log x$ yields a straight line only with equations in the form of Eq. (4-35).

## Gompertz Equation

The Gompertz equation,[4] or $S$ curve (Fig. 4-11), appears frequently in engineering practice. The Gompertz curve, for example, represents the sales volume vs. years for many products which have low sales when first introduced, experience a period of rapid increase, then reach saturation. The personnel required in many projects also often follows the curve. The form that represents Fig. 4-11 is

$$y = ab^{c^x} \tag{4-44}$$

where $a$, $b$, and $c$ are constants and $b$ and $c$ have magnitudes less than unity.

## Combination of Forms

It may be possible to fit a curve by combining two or more forms. For example, in Fig. 4-12, suppose that the value of $y$ approaches asymptotically a straight line

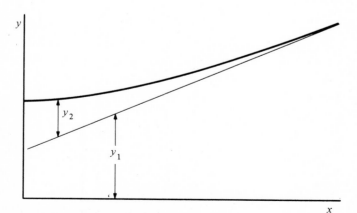

**Figure 4-12** Combination of two forms.

as $x$ increases. A reasonable way to attack this modeling task would be to propose that

$$y = y_1 + y_2 = (a + bx) + (c + dx^m)$$

where $m$ is a negative exponent.

## 4-13 AN OVERVIEW OF EQUATION FITTING

The task of finding suitable equations to represent the performance of components or thermodynamic properties is a common preliminary step to simulating and optimizing complex systems. Data may be available in tabular or graphic form, and we seek to represent the data with an equation that is both simple and faithful. A requirement for keeping the equation simple is to choose the proper terms (exponential, polynomial, etc.) to include in the equation. It is possible, of course, to include all the terms that could possibly be imagined, evaluate the coefficients by the method of least squares, and then eliminate terms that provide little contribution. This process is essentially one of *regression analysis*,[5] which also is used to assess which variables are important in representing the dependent variable.

The field of statistical analysis of data is an extensive one, and this chapter has only scratched the surface. On the other hand, much of the effort in the statistical analysis of data is directed toward fitting experimental data to equations where random experimental error occurs. In equation fitting for the design of thermal systems, since catalog tables and charts are the most frequent source of data, there usually has already been a process of smoothing of the experimental data coming from the laboratory. Because of the growing need for fitting catalog data to equations, many designers hope that manufacturers will present the equation that represents the table or graph to save each engineer the effort of developing the equation again when needed.

This chapter presented one approach to mathematical modeling where the relationship of dependent and independent variables was developed without the help of physical laws. Chapter 5 explores some special important cases where physical insight into some thermal equipment can be used to advantage in fitting equations to performance data.

## PROBLEMS

**4-1** Compute

$$\begin{vmatrix} 2 & -1 & 0 & 3 \\ 1 & -2 & 2 & 4 \\ -3 & 1 & 0 & -1 \\ 4 & 2 & 0 & 3 \end{vmatrix}$$

**Ans.:** 50

**4-2** Test the coefficient matrix in the set of linear equations

$$\begin{bmatrix} 1 & 2 & -2 & 3 \\ 2 & -1 & 3 & -2 \\ -1 & 3 & 1 & -4 \\ 1 & -3 & 5 & -5 \end{bmatrix} \begin{bmatrix} x_1 \\ x_2 \\ x_3 \\ x_4 \end{bmatrix} = \begin{bmatrix} 5 \\ 18 \\ -6 \\ 13 \end{bmatrix}$$

and determine whether the set of equations is dependent or independent.

**4-3** Using a computer program (gaussian elimination or any other that is available for solving a set of linear simultaneous equations), solve for the $x$'s:

$$\begin{aligned}
2x_1 + x_2 - 4x_3 + 6x_4 + 3x_5 - 2x_6 &= 16 \\
-x_1 + 2x_2 + 3x_3 + 5x_4 - 2x_5 &= -7 \\
x_1 - 2x_2 - 5x_3 + 3x_4 + 2x_5 + x_6 &= 1 \\
4x_1 + 3x_2 - 2x_3 + 2x_4 + x_6 &= -1 \\
3x_1 + x_2 - x_3 + 4x_4 + 3x_5 + 6x_6 &= -11 \\
5x_1 + 2x_2 - 2x_3 + 3x_4 + x_5 + x_6 &= 5
\end{aligned}$$

Ans.: 2, -1, 1, 0, 3, -4.

**4-4** A second-degree equation of the form

$$y = a + bx + cx^2$$

has been proposed to pass through the three $(x, y)$ points $(1, 3)$, $(2, 4)$, and $(2, 6)$. Proceed with the solution for $a$, $b$, and $c$.

(*a*) Describe any unusual problems encountered.

(*b*) Propose an alternate second-degree relation between $x$ and $y$ that will successfully represent these three points.

**4-5** Use data from Table 4-3 at $t = 0, 50$, and $100°C$ to establish a second-degree polynomial that fits $h_g$ to $t$. Using the equation, compute $h_g$ at $80°C$.

Ans.: 2643.3 kJ/kg.

**4-6** Using the data from Table 4-3 for $v_g$ at $t = 40, 60, 80$ and $100°C$, develop a third-degree equation similar in form to Eq. (4-16). Compute $v_g$ at $70°C$ using this equation.

Ans.: 4.91 m$^3$/kg.

## Table 4-3 Properties of saturated water

| Temperature | | Pressure | Specific volume | Enthalpy | |
|---|---|---|---|---|---|
| $t, °C$ | $T, K$ | $p$, kPa | $v_g$, m$^3$/kg | $h_f$, kJ/kg | $h_g$, kJ/kg |
| 0 | 273.15 | 0.6108 | 206.3 | -0.04 | 2501.6 |
| 10 | 283.15 | 1.227 | 106.4 | 41.99 | 2519.9 |
| 20 | 293.15 | 2.337 | 57.84 | 83.86 | 2538.2 |
| 30 | 303.15 | 4.241 | 32.93 | 125.66 | 2556.4 |
| 40 | 313.15 | 7.375 | 19.55 | 167.45 | 2574.4 |
| 50 | 323.15 | 12.335 | 12.05 | 209.26 | 2592.2 |
| 60 | 333.15 | 19.92 | 7.679 | 251.09 | 2609.7 |
| 70 | 343.15 | 31.16 | 5.046 | 292.97 | 2626.9 |
| 80 | 353.15 | 47.36 | 3.409 | 334.92 | 2643.8 |
| 90 | 363.15 | 70.11 | 2.361 | 376.94 | 2660.1 |
| 100 | 373.15 | 101.33 | 1.673 | 419.06 | 2676.0 |

**4-7** Lagrange interpolation is to be used to represent the enthalpy of saturated air, $h_s$ kJ/kg, as a function of the temperature $t°C$. The pairs of $(h_s, t)$ values to be used as the basis are $(9.470, 0)$, $(29.34, 10)$, $(57.53, 20)$, and $(99.96, 30)$.

(a) Determine the values of the coefficients $c_1$ to $c_4$ in the equation for $h_s$.

(b) Calculate $h_s$ at $15°C$.

**Ans.:** (b) From tables 42.09 kJ/kg.

**4-8** An equation of the form

$$y - y_0 = a_1(x - 1) + a_2(x - 1)^2$$

is to fit the following three $(x, y)$ points: $(1, 4)$, $(2, 8)$, and $(3, 10)$. What are the values of $y_0, a_1,$ and $a_2$?

**Ans.:** $a_1 = 5$.

**4-9** The pumping capacity of a refrigerating compressor (and thus the capability for developing refrigerating capacity) is a function of the evaporating and condensing pressures. The refrigerating capacities in kilowatts of a certain reciprocating compressor at combinations of three different evaporating and condensing temperatures are shown in Table 4-4. Develop an equation similar to the form of Eq. (4-33), namely,

$$q_e = c_1 + c_2 t_e + c_3 t_e^2 + c_4 t_c + \cdots + c_9 t_e^2 t_c^2$$

**Ans.:** $c_1$ to $c_9$ are 239.51, 10.073, −0.10901, −3.4100, −0.0025000, −0.20300, 0.0082004, 0.0013000, −0.000080005.

**4-10** The data in Table 4-4 are to be fit to an equation using Lagrange interpolation with a form similar to Eq. (4-34). The variable $x$ corresponds to $t_e$, $y$ corresponds to $t_c$, and $z$ to $q_e$. Compute the coefficient $c_{23}$.

**Ans.:** −0.02026.

**4-11** The values of $c_1$ and $c_2$ are to be determined so that the curve represented by the equation $y = c_1/(c_2 + x)^2$ passes through the $(x, y)$ points $(2, 4)$ and $(3, 1)$. Find the *two* $c_1$-$c_2$ combinations.

**Ans.:** One value of $c_1$ is $\frac{4}{9}$.

**4-12** Using the graphical method for the form $y = b + ax^m$ described in Sec. 4-9, determine the equation that represents the following pairs of $(x, y)$ points: $(0.2, 26)$, $(0.5, 7)$, $(1, 2.8)$, $(2, 1.3)$, $(4, 0.79)$, $(6, 0.65)$, $(10, 0.58)$, $(15, 0.54)$.

**Ans.:** $y = 0.5 + 2.3x^{-1.5}$.

**4-13** A function $y$ is expected to be of the form $y = cx^m$ and the $xy$ data develop a straight

**Table 4-4  Refrigerating capacity $q_e$ kW**

| Evaporating temperature $t_e$, °C | Condensing temperature $t_c$, °C | | |
|---|---|---|---|
| | 25 | 35 | 45 |
| 0 | 152.7 | 117.1 | 81.0 |
| 5 | 182.9 | 141.9 | 101.3 |
| 10 | 215.4 | 170.7 | 126.5 |

line on log-log paper. The line passes through the $(x, y)$ points (100, 50) and (1000, 10). What are the values of $c$ and $m$?

Ans.: $c = 1250$.

**4-14** Compute the constants in the equation $y = a_0 + a_1x + a_2x^2$ to provide a best fit in the sense of least squares for the following $(x, y)$ points: (1, 9.8), (3, 13.0), (6, 9.1), and (8, 0.6).

Ans.: $6.424, 3.953, -0.585$.

**4-15** With the method of least squares, fit the enthalpy of saturated liquid $h_f$ by means of a cubic equation to the temperature $t$ in degrees Celsius using the 11 points on Table 4-3. Then compute the values of $h_f$ at the 11 points with the equation just developed.

Ans.: $h_f = -0.0037 + 4.2000t - 0.000505t^2 + 0.000003935t^3$

**4-16** A frequently used form of equation to relate saturation pressures to temperatures is

$$\ln p = A + \frac{B}{T}$$

where $p$ = saturation pressure, kPa
$\quad T$ = absolute temperature, K

With the method of least squares and the 11 points for Table 4-3, determine the values of $A$ and $B$ that give the best fit. Then compute the values of $p$ at the 11 points using the equation just developed.

Ans.: $\ln p = 18.60 - 5206.9/T$.

**4-17** The variable $z$ is to be expressed in an equation of the form

$$z = ax + by + cxy$$

The following data points are available, and a least-squares fit is desired:

| $z$ | $x$ | $y$ |
|------|-----|-----|
| 0.1 | 1 | 1 |
| −0.9 | 1 | 2 |
| 2.0 | 2 | 2 |
| −1.8 | 3 | 1 |

Determine the values of $a$, $b$, and $c$.

Ans.: $-2.0467, -0.9167$, and $1.8833$.

**4-18** Three points, $(x_1, y_1)$, $(x_2, y_2)$, and $(x_3, y_3)$, lie precisely on the straight line $y = a + bx$. If a least-squares best fit were applied to these three points to determine the values of $A$ and $B$ in the equation $y = A + Bx$, show that the process would indeed give $A = a$, and $B = b$.

**4-19** An equation is to be found that represents the function shown in Fig. 4-13. Since one single simple expression seems inadequate, propose that $y = f_1(x) + f_2(x)$. Suggest appropriate forms for $f_1$ and $f_2$ and sketch these functions.

**4-20** The enthalpy of a solution is a function of the temperature $t$ and the concentration $x$ and consists of straight lines at a constant temperature, as shown in Fig. 4-14. Develop an equation that accurately represents $h$ as a function of $x$ and $t$.

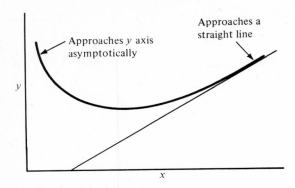

**Figure 4-13** Function in Prob. 4-19.

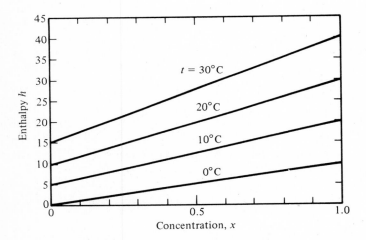

**Figure 4-14** Enthalpy as a function of temperature and concentration in Prob. 4-20.

# REFERENCES

1. *Procedures for Simulating the Performance of Components and Systems for Energy Calculations*, American Society of Heating, Refrigerating, and Air-Conditioning Engineers, New York, 1975.
2. M. W. Wambsganss, Jr., "Curve Fitting with Polynomials," *Mach. Des.*, vol. 35, no. 10, p. 167, Apr. 25, 1963.
3. C. Daniel and F. S. Wood, *Fitting Equations to Data*, Wiley-Interscience, New York, 1971.
4. D. S. Davis, *Nomography and Empirical Equations*, Reinhold, New York, 1955.
5. N. R. Draper and H. Smith, *Applied Regression Analysis*, Wiley, New York, 1966.

# FIVE

# MODELING THERMAL EQUIPMENT

## 5-1 USING PHYSICAL INSIGHT

This chapter continues the objective of the previous chapter, of fitting equations to the performance of components. Chapter 4, however, assumed that physical insight into the equation either did not exist or offered no particular advantage. In contrast, this chapter concentrates on three classes of components that appear almost universally in thermal systems, heat exchangers, distillation separators, and turbomachinery, where the knowledge of the physical relationships helps structure the equations.

The reasons for singling out these three components are different for each component. The preface of this book indicates that a background in heat transfer is assumed. In the study of thermal systems a segment of that knowledge is particularly important, namely, predicting the performance of an existing heat exchanger. Not only is the selection of a heat exchanger important, but it is also crucial to be able to calculate how a certain heat exchanger will perform when operating at off-design conditions. A useful tool to be stressed in this chapter is the *effectiveness* of heat exchangers.

The techniques this book explains are particularly applicable to the thermal processing industry, which includes petroleum refining, and other process industries where the separation of mixtures of several substances by means of distillation is an integral process. Even an understanding of the separation of binary mixtures which is explained in this chapter, expands the horizons of applications of the future topics of simulation and optimization.

The third class of thermal equipment treated in this chapter is turbomachinery, the performance of which can often be expressed in terms of dimensionless groups. This chapter will show how the use of dimensionless groups can sometimes simplify the equation representation of a turbomachine.

## 5-2 SELECTING VS. SIMULATING A HEAT EXCHANGER

A common engineering task is to *select*, design, or specify a heat exchanger to perform a certain heat-transfer duty. The engineer then decides on the type of heat exchanger and its details. Three of the several dozen types of heat exchangers available are shown in Fig. 5-1. Figure 5-1a shows a shell-and-tube heat exchanger, commonly used to transfer heat between two liquids. One of the fluids flows inside the tubes and is called the tube-side fluid, while the other flows over the outside of the tubes and is called the shell-side fluid. The heat exchanger in Fig. 5-1a has two tube passes, which means that the tube-side fluid

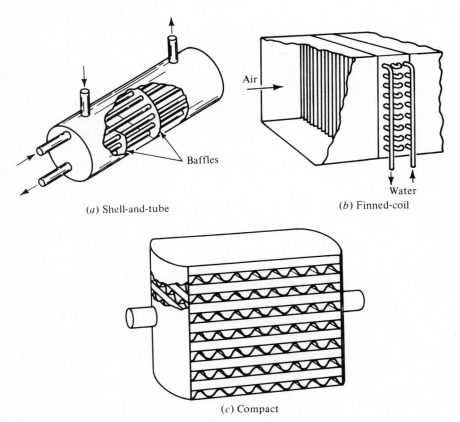

(a) Shell-and-tube

(b) Finned-coil

(c) Compact

**Figure 5-1**  Several types of heat exchangers.

flows through half the tubes in one direction and back through the other half. The head of the heat exchanger on the left end is equipped with a divider that separates the incoming from the outgoing tube-side fluid. Baffles are placed in the shell so that the shell-side fluid flows across the tubes a number of times before leaving the heat exchanger, instead of short-circuiting to the outlet.

The finned-coil heat exchanger in Fig. 5-1b is the type often chosen to transfer heat between a gas and a liquid. Since the resistance to heat transfer on the gas side is usually high because of the low heat-transfer coefficient of a gas, fins are installed on the gas side to increase the heat-transfer area.

The third type of heat exchanger, shown in Fig. 5-1c, is a compact heat exchanger; it usually consists of a stack of metal plates that are often corrugated and arranged so that the two fluids flow through alternate spaces between the plates.

We now return to the distinction between selecting and simulating a heat exchanger. To select a shell-and-tube heat exchanger, for example, the flow rates, entering temperatures, and leaving temperatures of both fluids would be known. The task of the designer is to select the combination of shell diameter, tube length, number of tubes, number of tube passes, and the baffle spacing that will accomplish the specified heat-transfer duty. The design must also ensure that certain pressure-drop limitations of the fluids flowing through the heat exchanger are not exceeded.

In simulation, on the other hand, the heat exchanger already exists, either in actual hardware or as a specific design. Furthermore the performance characteristics of the heat exchanger are available, such as the area and overall heat-transfer coefficients. Simulation of a heat exchanger consists of predicting outlet conditions, such as temperatures, for various inlet temperatures and flow rates. The emphasis of the next several sections will be on predicting outlet conditions of a given heat exchanger when the inlet conditions are known.

## 5-3 COUNTERFLOW HEAT EXCHANGER

For a heat exchange between two fluids with given inlet and outlet temperatures, the most favorable difference in temperature between these two fluids is achieved with a counterflow arrangement. A counterflow heat exchanger is therefore a good choice for a standard of comparison. Figure 5-2 shows a counterflow heat exchanger with the symbols that will be used in developing equations.

Three equations for the rate of heat transfer $q$ in watts are

$$q = W_1(t_{1,i} - t_{1,o}) \tag{5-1}$$

$$q = W_2(t_{2,o} - t_{2,i}) \tag{5-2}$$

$$q = UA \frac{(t_{1,i} - t_{2,o}) - (t_{1,o} - t_{2,i})}{\ln\left[(t_{1,i} - t_{2,o})/(t_{1,o} - t_{2,i})\right]} \tag{5-3}$$

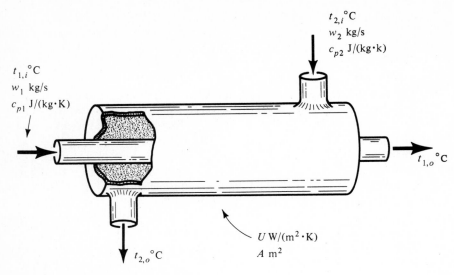

**Figure 5-2** A counterflow heat exchanger.

where $q$ = rate of heat transfer, W
$w_1$, $w_2$ = flow rates of respective fluids, kg/s
$c_{p1}$, $c_{p2}$ = specific heats of respective fluids, J/(kg · K)
$t$ = temperature, °C
$U$ = overall heat-transfer coefficient, W/(m² · K)
$A$ = heat-transfer area, m²
$W_1 = w_1 c_{p1}$
$W_2 = w_2 c_{p2}$

Equations (5-1) to (5-3) have made no assumption about which is the hot fluid and which the cold. If $q$ is positive, fluid 1 is hotter than fluid 2, but if $q$ is negative, the opposite is true. If the $W$'s, $UA$, and the entering temperatures are known, the three equations contain three unknowns, $q$, $t_{1,o}$, and $t_{2,o}$. The number of equations can be reduced to two by eliminating $q$, to give

$$W_1(t_{1,i} - t_{1,o}) = W_2(t_{2,o} - t_{2,i}) \tag{5-4}$$

$$W_1(t_{1,i} - t_{1,o}) = UA \frac{(t_{1,i} - t_{2,o}) - (t_{1,o} - t_{2,i})}{\ln\left[(t_{1,i} - t_{2,o})/(t_{1,o} - t_{2,i})\right]} \tag{5-5}$$

Solving for $t_{2,o}$ in Eq. (5-4) and substituting into Eq. (5-5) gives

$$\ln \frac{t_{1,i} - [t_{2,i} + (W_1/W_2)(t_{1,i} - t_{1,o})]}{t_{1,o} - t_{2,i}} = UA\left(\frac{1}{W_1} - \frac{1}{W_2}\right)$$

Define $D$ as

$$D = UA \left( \frac{1}{W_1} - \frac{1}{W_2} \right)$$

Then

$$\frac{t_{1,i} - t_{2,i} - (W_1/W_2)(t_{1,i} - t_{1,o})}{t_{1,o} - t_{2,i}} = e^D$$

Solving for $t_{1,o}$ gives

$$t_{1,o} = \frac{t_{1,i}(W_1/W_2 - 1) + t_{2,i}(1 - e^D)}{W_1/W_2 - e^D}$$

or, in alternate form,

$$t_{1,o} = t_{1,i} - (t_{1,i} - t_{2,i}) \frac{1 - e^D}{W_1/W_2 - e^D} \tag{5-6}$$

Equation (5-6) permits computation of one outlet temperature of a heat exchanger of known characteristics when both entering temperatures are known. The other outlet temperature can be computed by application of Eq. (5-4), and $q$ can be computed from Eq. (5-1). If the outlet temperature of fluid 2 is the one sought directly, subscripts 1 and 2 can be interchanged in Eq. (5-6) and in the equation for $D$.

## 5-4 SPECIAL CASE OF COUNTERFLOW HEAT EXCHANGER WHERE PRODUCTS OF FLOW RATES AND SPECIFIC HEATS ARE EQUAL

The direct application of Eq. (5-6) is unsuccessful when

$$w_1 c_{p1} = w_2 c_{p2} \qquad \text{thus} \qquad W_1 = W_2$$

The value of $D$ is zero, and Eq. (5-6) is indeterminate. There are two ways to develop an alternate expression for the outlet temperature, one mathematical and the other by physical reasoning.

The mathematical solution uses the expression for $e^x$ as a series,

$$1 + x + \frac{x^2}{2} + \cdots$$

The indeterminate part of Eq. (5-6) can be written

$$\frac{1 - e^D}{W_1/W_2 - e^D} = \frac{1 - \left\{ 1 + \frac{UA}{W_1} \left( 1 - \frac{W_1}{W_2} \right) + \frac{1}{2} \left[ \frac{UA}{W_1} \left( 1 - \frac{W_1}{W_2} \right) \right]^2 + \cdots \right\}}{\frac{W_1}{W_2} - \left\{ 1 + \frac{UA}{W_1} \left( 1 - \frac{W_1}{W_2} \right) + \frac{1}{2} \left[ \frac{UA}{W_1} \left( 1 - \frac{W_1}{W_2} \right) \right]^2 + \cdots \right\}}$$

Canceling where possible and dividing both the numerator and denominator by $1 - W_1/W_2$ results in

$$\frac{1 - e^D}{(W_1/W_2) - e^D} = \frac{-(UA/W_1) - \frac{1}{2}(UA/W_1)^2(1 - W_1/W_2) + \cdots}{-1 - UA/W_1 - \frac{1}{2}(UA/W_1)^2(1 - W_1/W_2) + \cdots}$$

Finally let $W_1 \rightarrow W_2$ and call this common value $W$. Then

$$\frac{1 - e^D}{W/W - e^D} = \frac{UA/W}{1 + UA/W} = \frac{1}{W/UA + 1}$$

Substituting back into Eq. (5-6) gives

$$t_{1,o} = t_{1,i} - \frac{t_{1,i} - t_{2,i}}{W/UA + 1} \tag{5-7}$$

which is the equation for computing one outlet temperature.

The physical analysis that leads to Eq. (5-7) is to recognize that the change of temperature of one fluid while flowing past a differential area $dA$ in Fig. 5-3 equals the change in temperature of the other fluid. The slopes of the two temperature lines are then the same at all positions along the area. (This condition does not yet establish that the lines are straight.) A further stipulation demanded by the heat-transfer rate equation is that

$$dq = W \, dt_2 = W \, dt_1 = U \, dA \, (t_1 - t_2) \tag{5-8}$$

Starting at the left edge of the graph in Fig. 5-3, the inlet temperature $t_{2,i}$ of fluid is specified, and there is some outlet temperature of fluid 1, designated $t_{1,o}$, yet to be determined. Regardless of the value of $t_{1,o}$, the slopes of the temperature lines are fixed for the first increment of area $dA$

$$\frac{dt_1}{dA} = \frac{dt_2}{dA} = \frac{U}{W}(t_{1,o} - t_{2,i})$$

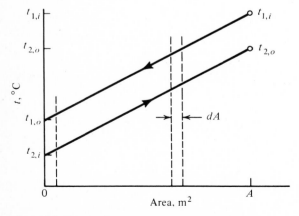

**Figure 5-3** Counterflow heat exchanger where $W_1 = W_2$.

Since the slopes through $dA_1$ are identical, the temperature difference after $dA$ is still $t_{1,o} - t_{2,i}$, and by similar reasoning the temperature difference remains constant through the entire heat exchanger. Thus the temperature lines are straight, parallel lines. The choice of $t_{1,o}$ was arbitrary, but now we see that it must be chosen so that the $t_1$ line terminates at the specified inlet temperature of $t_{1,i}$. Since the mean temperature difference in the heat exchanger is $t_{1,o} - t_{2,i}$, the rate equation can be written

$$q = UA(t_{1,o} - t_{2,i}) = W(t_{1,i} - t_{1,o})$$

Then
$$t_{1,o}(UA + W) = Wt_{1,i} + UAt_{2,i}$$

and
$$t_{1,o} = \frac{Wt_{1,i} + UAt_{2,i}}{UA + W}$$

Add zero in the form of $(UAt_{1,i} - UAt_{1,i})/(UA + W)$ to the right side; then

$$t_{1,o} = t_{1,i} - \frac{t_{1,i} - t_{2,i}}{W/UA + 1}$$

which checks the mathematical derivation, Eq. (5-7).

**Example 5-1** In the counterflow heat exchanger shown in Fig. 5-4, a flow rate of 0.5 kg/s of water enters one circuit of the heat exchanger at a temperature of 30°C, and the same flow rate of water enters the other circuit at a temperature of 65°C. The $UA$ of the heat exchanger is 4 kW/K. What is the mean temperature difference between the two streams?

SOLUTION The products of the flow rates and specific heats of the two streams are the same, $(0.5 \text{ kg/s}) [4.19 \text{ kJ/(kg·K)}] = 2.095 \text{ kW/K}$. For this special case where the $wc_p$ products are equal, Eq. (5-7) gives the outlet temperature for the hot stream $t_{1,o}$

$$t_{1,o} = 65 - \frac{65 - 30}{2.095/4 + 1} = 42°C$$

The temperature difference at either end of the heat exchanger prevails throughout, and so the mean temperature difference is $42 - 30 = 12°C$.

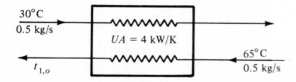

**Figure 5-4** Heat exchanger in Example 5-1.

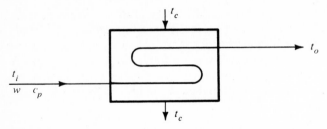

**Figure 5-5** An evaporator or condenser where one fluid remains at a constant temperature.

## 5-5 EVAPORATORS AND CONDENSERS

A special set of equations is possible—and indeed necessary—when one of the fluids flowing through a heat exchanger changes phase. In an evaporator or condenser, as shown in Fig. 5-5, assume that there is no superheating or subcooling of the fluid that changes phase. That fluid will then remain at a constant temperature, provided that its pressure does not change.

The log-mean temperature difference still applies and in combination with a heat balance gives

$$q = UA \frac{(t_c - t_i) - (t_c - t_o)}{\ln\,[(t_c - t_i)/(t_c - t_o)]} = wc_p(t_o - t_i) \tag{5-9}$$

Equation (5-9) can be converted into the form

$$\frac{UA}{wc_p} = \ln \frac{t_c - t_i}{t_c - t_o} = -\ln \frac{t_c - t_o}{t_c - t_i}$$

Taking the antilog gives

$$e^{-UA/wc_p} = \frac{t_c - t_o}{t_c - t_i} = \frac{t_c - t_o + t_i - t_i}{t_c - t_i}$$

Then

$$t_o = t_i + (t_c - t_i)(1 - e^{-UA/wc_p}) \tag{5-10}$$

For a heat exchanger of known characteristics Eq. (5-10) can be used to compute the outlet temperature of the fluid that does not change phase when its entering temperature and the temperature of the boiling or condensing fluid $t_c$ are known.

The characteristic shape of the temperature curves of the two fluids is shown in Fig. 5-6, applicable to a condenser.

**Example 5-2** Water is continuously heated from 25 to 50°C by steam condensing at 110°C. If the water flow rate remains constant but its inlet temperature drops to 15°C, what will its new outlet temperature be?

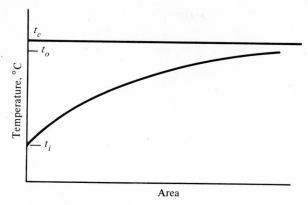

**Figure 5-6** Temperature distribution in fluids in a condenser.

SOLUTION  The terms $U, A, w, c_p$, and $t_c$ all remain constant, so

Original             New

$$\frac{t_o - t_i}{t_c - t_i} = 1 - e^{-UA/wc_p} = \frac{t_o - t_i}{t_c - t_i}$$

$$\frac{50 - 25}{110 - 25} = \frac{t_o - 15}{110 - 15}$$

New $t_o = 42.9°\text{C}$.

## 5-6 HEAT-EXCHANGER EFFECTIVENESS

The effectiveness $\epsilon$ of a heat exchanger is defined as

$$\epsilon = \frac{q_{\text{actual}}}{q_{\text{max}}} \tag{5-11}$$

where $q_{\text{actual}}$ = actual rate of heat transfer, kW

$q_{\text{max}}$ = maximum possible rate of heat transfer, kW, with same inlet temperatures, flow rates, and specific heats as actual case

Another approach to defining $q_{\text{max}}$ is to designate it as the rate of heat transfer that a heat exchanger of infinite area would transfer with given inlet temperatures, flow rates, and specific heats.

**Example 5-3** What is the maximum rate of heat transfer possible in a counterflow heat exchanger shown in Fig. 5-7 if water enters at 30°C and cools oil entering at 60°C?

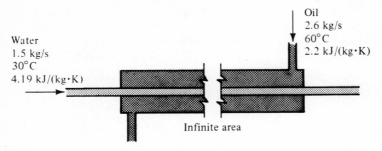

**Figure 5-7** Oil cooler in Example 5-3.

$$\text{Flow rate} = \begin{cases} 2.6 \text{ kg/s} & \text{oil} \\ 1.5 \text{ kg/s} & \text{water} \end{cases}$$

$$\text{Specific heat} = \begin{cases} 2.2 \text{ kJ/(kg} \cdot \text{K)} & \text{oil} \\ 4.19 \text{ kJ/(kg} \cdot \text{K)} & \text{water} \end{cases}$$

SOLUTION The break in the heat exchanger indicates that to achieve the maximum rate of heat transfer the area must be made infinite. The next question, then, is: What are the outlet temperatures? Does the oil leave at 30°C, or does the water leave at 60°C?

From energy balances those two options give the following consequences:

1. Oil leaves at 30°C

$$q = (2.6 \text{ kg/s}) \, [2.2 \text{ kJ/(kg} \cdot \text{K)}] \, (60 - 30°\text{C}) = 171.6 \text{ kW}$$

and water leaves at

$$30°\text{C} + \frac{171.6 \text{ kW}}{(1.5 \text{ kg/s}) \, [4.19 \text{ kJ/(kg} \cdot \text{K)}]} = 57.3°\text{C}$$

2. Water leaves at 60°C

$$q = (1.5)(4.19)(60 - 30) = 188.6 \text{ kW}$$

and oil leaves at

$$60°\text{C} - \frac{188.6}{(2.6)(2.2)} = 27°\text{C}$$

The second case is clearly impossible because the oil temperature would drop below that of the entering water, which would violate the second law of thermodynamics. Thus, $q_{max} = 171.6$ kW.

The concept that Example 5-3 has exposed is that the maximum rate of heat transfer occurs when the fluid with the minimum product of flow rate and

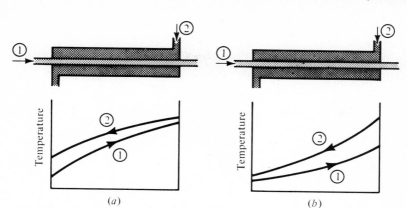

**Figure 5-8** Temperature profiles in a counterflow heat exchanger.

specific heat changes temperature to the entering temperature of the other fluid. Equation (5-11) can be rewritten

$$\epsilon = \frac{q_{\text{actual}}}{(wc_p)_{\text{min}}(t_{\text{hot,in}} - t_{\text{cold,in}})} \tag{5-12}$$

where $(wc_p)_{\text{min}}$ is the smaller $wc_p$ of the two fluids.

## 5-7 TEMPERATURE PROFILES

The shapes of the temperature curves in the counterflow heat exchanger are shown by Fig. 5-8a or b. The minimum product of $wc_p$ is possessed by fluid 1 in Fig. 5-8a and by fluid 2 in Fig. 5-8b, which is indicated by the fluid whose temperature changes most. The curves are steepest in the portion of the heat exchanger where the rate of heat transfer is highest, and this region occurs where the temperature differences are largest.

## 5-8 EFFECTIVENESS OF A COUNTERFLOW HEAT EXCHANGER

The form of Eq. (5-12) can be applied to the equation for a counterflow heat exchanger, Eq. (5-6), to develop an expression for the effectiveness of the counterflow heat exchanger. Denote fluid 1 as the fluid with the lesser value of $wc_p$. Note that

$$\frac{t_{1,o} - t_{1,i}}{t_{2,i} - t_{1,i}} = \frac{(wc_p)_{\text{min}}(t_{1,o} - t_{1,i})}{(wc_p)_{\text{min}}(t_{2,i} - t_{1,i})}$$

$$= \frac{q_{\text{actual}}}{(wc_p)_{\text{min}}(t_{\text{hot,in}} - t_{\text{cold,in}})} = \epsilon$$

Combining with Eq. (5-6) gives

$$\epsilon = \frac{t_{1,o} - t_{1,i}}{t_{2,i} - t_{1,i}} = \frac{1 - e^D}{W_{min}/W_2 - e^D} \tag{5-13}$$

where

$$D = UA \left( \frac{1}{W_{min}} - \frac{1}{W_2} \right) = \frac{UA}{W_{min}} \left( 1 - \frac{W_{min}}{W_2} \right) \tag{5-14}$$

Examination of Eqs. (5-13) and (5-14) shows that the effectiveness of the counterflow heat exchanger can be expressed as a function of two dimensionless groups, $UA/W_{min}$ and $W_{min}/W_2$.

## 5-9 NUMBER OF TRANSFER UNITS, NTU'S

The ability to express the effectiveness as a function of $UA/W_{min}$ and $W_{min}/W_2$ leads to graphic presentations of effectiveness of heat exchangers of various configurations. The group $UA/W_{min}$ is called the *number of transfer units* (NTU). Typical graphic presentations are shown in Figs. 5-9 and 5-10 for two different configurations of heat exchangers. Figure 5-9 is the graphic presentation of Eqs. (5-13) and (5-14).

> **Example 5-4** Compute the effectiveness of a counterflow heat exchanger having a $U$ value of 1.1 kW/(m² · K) and an area of 16 m² when one fluid has a flow rate of 6 kg/s and a specific heat of 4.1 kJ/(kg · K) and the other fluid a flow rate of 3.8 kg/s and a specific heat of 3.3 kJ/(kg · K).

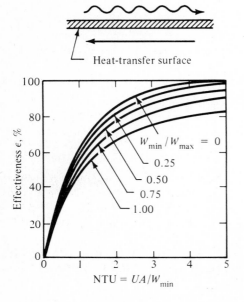

**Figure 5-9** Effectiveness of a counter-flow heat exchanger. (*From W. M. Kays and A. L. London, Compact Heat Exchangers, 2d ed., McGraw-Hill Book Company, New York, 1964, p. 50; used by permission.*)

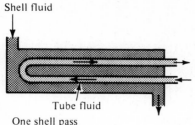

Shell fluid

Tube fluid

One shell pass
2, 4, 6, . . . tube passes

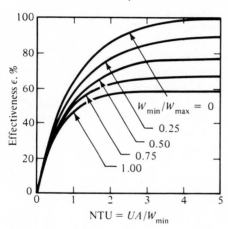

**Figure 5-10** Effectiveness of a parallel counterflow heat exchanger. (*From W. M. Kays and A. L. London, Compact Heat Exchangers, 2d ed., McGraw-Hill Book Company, New York, 1964, p. 54; used by permission.*)

### SOLUTION

$$W_a = (6 \text{ kg/s}) [4.1 \text{ kJ/(kg} \cdot \text{K)}] = 24.6 \text{ kW/K}$$

$$W_b = (3.8)(3.3) = 12.54 \text{ kW/K} \qquad \text{designate } W_{\min}$$

$$\text{NTU} = \frac{UA}{W_{\min}} = \frac{(1.1)(16)}{12.54} = 1.40$$

$$\frac{W_{\min}}{W_2} = \frac{12.54}{24.6} = 0.51$$

From Fig. 5-9

$$\epsilon = 0.67$$

or from Eq. (5-14)

$$D = 1.40(1 - 0.51) = 0.686$$

and from Eq. (5-13)

$$\epsilon = \frac{1 - e^{0.686}}{0.51 - e^{0.686}} = 0.668$$

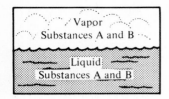

Figure 5-11 A binary solution.

It is now appropriate to put the previous nine sections on modeling of heat exchangers into perspective. The direction from which the modeling was approached was to predict the performance of an existing heat exchanger, which is a common assignment in systems simulation. The outlet temperatures can be computed if the inlet temperatures and effectiveness are known. The effectiveness can be computed or determined from a graph when the product of $UA$ and the products of mass rate of flow and specific heat are known. Equations for the effectiveness of a counterflow heat exchanger have been presented, and graphs are available[1] for various configurations of heat exchangers. Worsøe-Schmidt[2] and Høgaard Knudsen have provided coefficients for equations that fit the curves of Fig. 5-10 as well as other heat exchangers.

## 5-10 BINARY SOLUTIONS

Figure 5-11 illustrates a binary solution where equilibrium prevails between liquid and vapor. The liquid is a solution of substances A and B, both exerting a vapor pressure such that both substances A and B exist in the vapor. The concentration of substance A in the vapor phase is not necessarily the same as its concentration in the liquid phase. A frequent assignment in petroleum, cryogenic, petrochemical, food, and other thermal processing industries is to separate substance A from substance B. System simulation computer programs (Chap. 6) are heavily used in these industries, and the separation process (usually by distillation) of binary solutions is often an integral function of these programs. We wish to be able to incorporate distillation equipment into systems along with heat exchangers, fans, pumps, compressors, expanders, and other thermal components.

## 5-11 TEMPERATURE-CONCENTRATION-PRESSURE CHARACTERISTICS

The temperature-concentration, $t$-vs.-$x$, graph of a binary mixture for a given pressure is shown in Fig. 5-12. The abscissa is the fraction of material A and a complementary scale is the fraction of material B. The fractions can be expressed either as mass fractions, for example,

$$\text{Mass fraction of A} = \frac{\text{mass of A, kg}}{\text{mass of A, kg} + \text{mass of B, kg}}$$

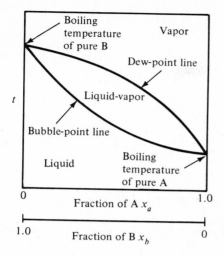

**Figure 5-12** Temperature-concentration diagram of a binary solution at a constant pressure.

or as mole fractions

$$\text{mole fraction of A} = \frac{\text{mol A}}{\text{mol A} + \text{mol B}}$$

$$= \frac{(\text{mass of A, kg})/(MW_a)}{(\text{mass of A, kg})/(MW_a) + (\text{mass of B, kg})/(MW_b)}$$

where $MW_a$ and $MW_b$ are the molecular weights of substances A and B, respectively.

The curves on Fig. 5-12 mark off three regions on the graph. The upper region is vapor (actually superheated vapor). If the temperature of vapor is reduced at a constant concentration, the state reaches the dew-point line, where some vapor begins to condense. The region between the two curves is the liquid-vapor region, where both vapor and liquid exist in equilibrium, as represented by the vessel in Fig. 5-11. The lowest region in Fig. 5-12 is the liquid region

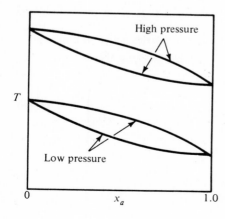

**Figure 5-13** $T$ vs. $x$ diagrams for two different pressures.

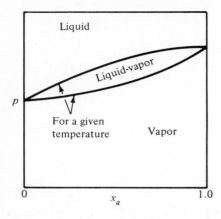

**Figure 5-14** Pressure-concentration diagram.

(actually subcooled liquid). If the temperature of subcooled liquid is increased at a constant concentration, the state reaches the bubble-point line, where some liquid begins to vaporize.

The temperature where the curves join at the left axis is the boiling temperature of substance B at the pressure in question, and the temperature intercepted at the right axis is the boiling temperature of substance A.

The pair of curves shown in Fig. 5-12 applies to a given pressure, and there will be a pair of curves for each different pressure, as shown in Fig. 5-13.

An alternate form of expressing the properties of a binary solution is to present a $p$-vs.-$x$ diagram, as in Fig. 5-14, where the pair of curves apply to a given temperature.

## 5-12 DEVELOPING A $T$-VS.-$x$ DIAGRAM

Several ideal relationships can be assembled to compute the data for a graph like Fig. 5-12. In some cases the results will be very close to the actual binary properties, while with other combinations of substances the real properties deviate from the ideal. In most cases the idealization will give a fair approximation of the properties of the real mixture.

The three tools used to develop the binary properties are (1) the saturation pressure-temperature relationships of the two substances, (2) Raoult's law, and (3) Dalton's law.

**Saturation pressure-temperature relation** A simple equation form that relates pressure and temperature at saturated conditions (see Prob. 4-16) is

$$\ln P = C + \frac{D}{T}$$

where $P$ = saturation pressure, kPa
$T$ = absolute temperature, K
$C, D$ = constants

The values of $C$ and $D$ are unique to each substance and must be developed from experimental data. It is presumed that $C_a$ and $D_a$ for substance A and $C_b$ and $D_b$ for substance B are known in the equations

$$\ln P_a = C_a + \frac{D_a}{T} \tag{5-15}$$

$$\ln P_b = C_b + \frac{D_b}{T} \tag{5-16}$$

**Raoult's law** Raoult's law states that the vapor pressure of one component in a mixture is

$$p_a = x_{a,l} P_a \tag{5-17}$$

where $p_a$ = vapor pressure of substance A in mixture, kPa
$x_{a,l}$ = mole fraction of substance A in liquid, dimensionless
$P_a$ = saturation pressure of pure A at existing temperature, kPa

Thus, as for the binary mixture of A and B in Fig. 5-15, if at the temperature $T$ the saturation pressure of pure A is 500 kPa and the mole fraction in the liquid $x_{a,l} = 0.3$, the partial pressure of substance A in the vapor is $(0.3)(500) = 150$ kPa.

**Dalton's law** Dalton's law states that the total pressure of the vapor mixture is the sum of the partial pressures of the constituents

$$p_t = p_a + p_b \tag{5-18}$$

Furthermore,
$$p_a = x_{a,v} p_t \tag{5-19}$$

and
$$p_b = x_{b,v} p_t \tag{5-20}$$

where $p_t$ = total pressure, kPa
$x_{a,v}$ = mole fraction of A in vapor
$x_{b,v}$ = mole fraction of B in vapor

**Example 5-5** A binary solution of $n$-butane and $n$-heptane exists in liquid-vapor equilibrium at a pressure of 700 kPa. The saturation pressure-temperature relationships are

Butane: $$\ln P = 21.77 - \frac{2795}{T}$$

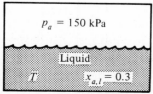

$p_a = 150$ kPa

Liquid

$T$    $x_{a,l} = 0.3$

$P_a$ at $T = 500$ kPa

**Figure 5-15** Raoult's law.

Heptane: $$\ln P = 22.16 - \frac{3949}{T}$$

where $P$ is in pascals. Compute the mole fraction of butane in the liquid and in the vapor at a temperature of 120°C.

SOLUTION At 120°C the saturation pressures of pure substances are

Butane: $$P_{but} = \exp\left(21.77 - \frac{2795}{393}\right) = 2322 \text{ kPa}$$

Heptane: $$P_{hep} = \exp\left(22.16 - \frac{3949}{393}\right) = 182 \text{ kPa}$$

The combination of Eqs. (5-17) and (5-18) along with the recognition that $x_{hep,l} + x_{but,l} = 1.0$ yields

$$p_{hep} + p_{but} = (1 - x_{but,l})(182) + (x_{but,l})(2322) = 700$$

and so $$x_{but,l} = 0.242$$

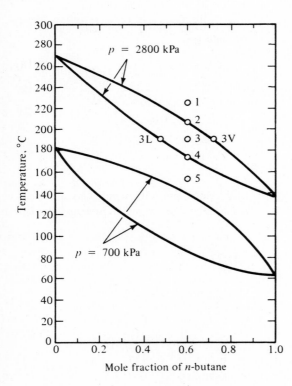

Figure 5-16 Binary system of n-butane and n-heptane.

To find the concentration of butane in the vapor, use Eqs. (5-18) to (5-20)

$$x_{but,v} = \frac{x_{but,l}P_{but}}{p_t} = \frac{(0.242)(2322)}{700} = 0.796$$

The results may be compared with the experimentally determined properties shown in Fig. 5-16.

## 5-13 CONDENSATION OF A BINARY MIXTURE

When a pure substance condenses at constant pressure, the temperature remains constant but in a binary mixture the temperature progressively changes, even though the pressure remains constant. The condensation process of a binary system taking place in a tube (Fig. 5-17) can be represented for a specific binary mixture as in Fig. 5-16. Assume that the condensation takes place at a constant pressure of 2800 kPa and vapor enters with a mole fraction of 0.6 $n$-butane. Point 1 is superheated vapor at a temperature of 223°C, which is first cooled until point 2, where the temperature is 204°C and condensation begins. Removal of heat from the mixture results in continued condensation. At point 3 the temperature has dropped to 190°C, and here the liquid has a concentration of 0.48 and the vapor 0.71. The system is a mixture of liquid and vapor at point 3, and a mass balance of the butane can indicate how much liquid and how much vapor exists at 3. For 1 mol of combination

0.48 mol in liquid state + (0.71)(1 − mol in liquid state) = (1 mol)(0.60)

Therefore at point 3 in 1 mol of combination there is 0.478 mol of liquid and 0.522 mol of vapor.

Condensation continues to point 4, where the temperature is 173°C; here all the vapor has condensed. Further removal of heat results in subcooling the liquid. During the condensation process the temperature drops from 204 to173°C.

## 5-14 SINGLE-STAGE DISTILLATION

A simple distillation unit is the single-stage still shown in Fig. 5-18. If the still operates at a pressure for which the bubble-point and dew-point curves are shown in Fig. 5-19, and if the entering liquid to the partial vaporizer is at point

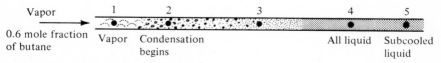

Vapor

0.6 mole fraction of butane    Vapor    Condensation begins      All liquid    Subcooled liquid

**Figure 5-17** Condensation of a binary mixture.

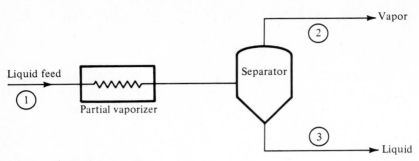

**Figure 5-18** Single-stage still.

1, various outlet conditions of the vapor and liquid are possible. The limiting cases are combination 2-3 and combination 2″-3″. If the liquid is heated to point 3, only liquid would leave the still and no vapor. If the vaporizer carries on the process to 2″, only vapor leaves the still and no liquid. The desired operating condition will be 2′-3′, for example, where there is some separation in the sense that the vapor leaves with a high concentration of A and the liquid leaves with a lower concentration than entered at 1.

> **Example 5-6** A single-stage distillation tower receives 3.2 mol/s of butane-heptane (Fig. 5-16). Liquid enters with a mole fraction of 0.4 butane, the still operates at 700 kPa, and the mixture leaves the partial vaporizer at a temperature of 120°C. What are the flow rates of liquid and vapor leaving the separator?
>
> SOLUTION At 120°C and a pressure of 700 kPa, the mole fraction of butane in the liquid is 0.29 and in the vapor is 0.75. A material balance on the butane states that
>
> $$(3.2 \text{ mol/s})(0.4) = (w_l \text{ mol/s})(0.29) + (3.2 - w_l)(0.75)$$
>
> $$w_l = 2.43 \text{ mol/s} \qquad w_v = 3.2 - 2.43 = 0.77 \text{ mol/s}$$

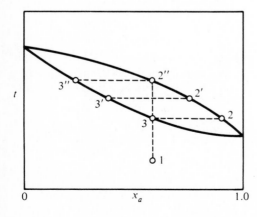

**Figure 5-19** Some possible outlet conditions from the still in Fig. 5-18.

## 5-15 RECTIFICATION

The single-stage still described in Sec. 5-14 performs a separation but has rather poor performance. The other end of the performance spectrum is an ideal rectification tower, shown in Fig. 5-20. The $t$-vs.-$x$ properties corresponding to the key locations in the rectification column are shown in Fig. 5-21. Liquid feed, assumed saturated, enters at point $i$, passes through a partial vaporizer, and is heated to temperature 2. In the column, vapor flows upward and liquid downward. The liquid is heated at the bottom to drive off some vapor, and a heat exchanger at the top of the tower condenses some of the vapor, providing a source of liquid to drain down the column.

Figure 5-21 shows that the separation of the two components is quite effective, since the liquid leaves the bottom of the tower at condition 3L and vapor leaves the top at condition 1V. There is a continuous transfer of heat and mass between the rising vapor and descending liquid. Furthermore in the ideal rectification tower there is equilibrium of temperature and vapor pressure between the liquid and vapor along the column. For accurate simulation of towers there must be a temperature difference between the vapor and liquid to transfer heat and vapor-pressure difference between the vapor and liquid in order to transfer mass.[3-6]

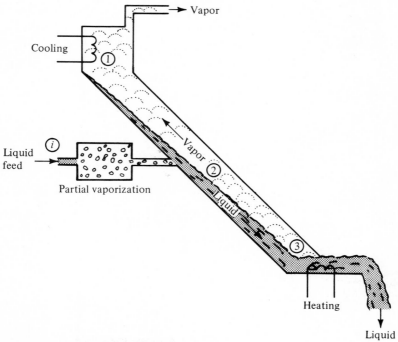

**Figure 5-20** A rectification column.

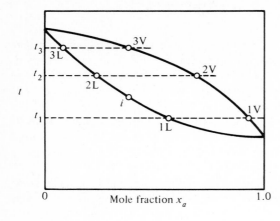

**Figure 5-21** States of binary system in rectification column.

## 5-16 ENTHALPY

Enthalpy values of binary solutions and mixtures of vapor are necessary when making energy calculations. For system simulation (Chap. 6) the enthalpy data would be most convenient in equation form. More frequently the enthalpy data appear in graphic form, as shown in the skeleton diagram of Fig. 5-22. Figure 5-22 is an enthalpy-concentration, $h$-vs.-$x$, diagram for solutions and vapor. Since pressure has a negligible effect on the enthalpy of the liquid, the chart is applicable to subcooled as well as saturated liquid, but because the enthalpy of the vapor is somewhat sensitive to the pressure, the enthalpy curves for vapor apply only to saturated conditions.

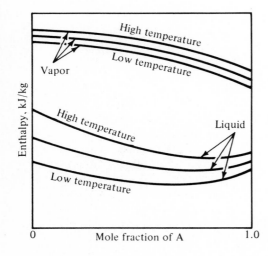

**Figure 5-22** Form of an enthalpy-concentration diagram.

## 5-17 PRESSURE DROP AND PUMPING POWER

A cost that appears in most economic analyses of thermal systems is the pumping cost. The size of a heat exchanger transferring heat to a liquid can be reduced, for example, if the flow rate of liquid or the velocity for a given flow rate is increased. The cost whose increase eventually overtakes the reduction in the cost of the heat exchanger as the velocity or flow increase is the pumping cost. Another example of the emergence of pumping cost is in the selection of optimum pipe size. The smaller the pipe the less the first cost but (for a given flow) the higher the pumping cost for the life of the system.

Since the pumping-cost term appears so frequently, it is appropriate to review the expression for pumping power. The pressure drop of an incompressible fluid flowing turbulently through pipes, fittings, heat exchangers, and almost any confining conduit varies as $w^n$

$$\Delta p = C(w^n)$$

where $C$ is a constant, $w$ is the mass rate of flow, and the exponent $n$ varies between about 1.8 and 2.0. Generally the value of $n$ is close to 2.0, except for flow in straight pipes at Reynolds numbers in the low turbulent range.

The ideal work per unit mass required for pumping fluid in steady flow is $\int v\, dp$, and for an incompressible fluid the power required is

$$\text{Power} = \frac{w}{\rho} \Delta p$$

where $\rho$ is the density.

Thus the equation for the pumping power is

$$\text{Power} = \frac{C}{\rho} w^{n+1} \approx \frac{C}{\rho} w^3 \qquad (5\text{-}21)$$

which is further modified by dividing by the pump, fan, or compressor efficiency.

## 5-18 TURBOMACHINERY

The methods of mathematical modeling explained in this chapter have generally been limited to expressing one variable as a function of one or two other variables. In principle it is possible to extend these methods to functions of three variables, but the execution might be formidable. Turbomachines, such as fans, pumps, compressors, and turbines, are used in practically all thermal systems, and in these components the dependent variable may be a function of three or more independent variables. Fortunately, the tool of dimensional analysis frequently permits reducing the number of independent variables to a smaller number by treating groups of terms as individual variables. The performance

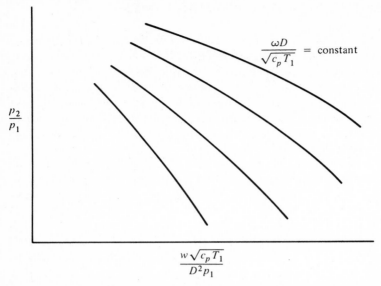

**Figure 5-23** Performance of a centrifugal compressor expressed in dimensionless groups to reduce the number of independent variables.

of a centrifugal compressor, for example, will typically appear as in Fig. 5-23. Instead of attempting to express $p_2$ as a function of six variables, an equation could be developed to express $p_2/p_1$ as a function of the two other dimensionless groups. Then in calculating $p_2$ the two independent variables $\omega D/\sqrt{c_p T_1}$ and $w\sqrt{c_p T_1}/D^2 p_1$ would be calculated first and next $p_2/p_1$; finally, $p_2$ would be computed from $p_2/p_1$ and $p_1$.

## PROBLEMS

**5-1** What is the effectiveness of a counterflow heat exchanger that has a $UA$ value of 24 kW/K if the respective mass rates of flow and specific heats of the two fluids are 10 kg/s, 2 kJ/(kg · K) and 4 kg/s, 4kJ/(kg · K)?

    **Ans.: 0.636.**

**5-2** Water flows through one side of a heat exchanger with a flow rate of 0.2 kg/s rising in temperature from 20 to 50°C. The specific heat of water is 4.19 kJ/(kg · K). The fluid on the other side of the heat exchanger enters at 80°C and leaves at 40°C. What is the effectiveness of the heat exchanger?

    **Ans.: 0.667.**

**5-3** A flow rate of 2 kg/s of water, $c_p$ = 4.19 kJ/(kg · K), enters one end of a counterflow heat exchanger at a temperature of 20°C and leaves at 40°C. Oil enters the other side of the heat exchanger at 60°C and leaves at 30°C. If the heat exchanger were made infinitely large while the entering temperatures and flow rates of the water and oil remained constant, what would the rate of heat transfer in the exchanger be?

    **Ans.: 223.5 kW.**

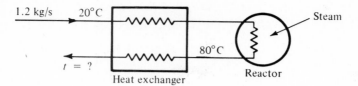

**Figure 5-24** Regenerative heat exchanger in Prob. 5-6.

**5-4** A flow rate of 0.8 kg/s of water is heated in a heat exchanger by condensing steam at 100°C. When water enters at 15°C, it leaves the heat exchanger at 62°C. If the inlet water temperature were changed to 20°C while its flow rate and the condensing temperature remained constant, what would its outlet temperature be?

   Ans.: 64.2°C.

**5-5** A counterflow heat exchanger cools 5 kg/s of oil, $c_p$ = 2.4 kJ/(kg · K), with water that has a flow rate of 7.5 kg/s. The specific heat of water is 4.19 kJ/(kg · K). Under the original operating conditions the oil is cooled from 75 to 40°C when water enters at 25°C. To what temperature will the oil be cooled if it enters at 65°C and if there is no change in the entering water temperature, the flow rates of either fluid, or the heat-transfer coefficients?

   Ans.: 37°C.

**5-6** In a processing plant a material must be heated from 20 to 80°C in order for the desired reaction to proceed, whereupon the material is cooled in a regenerative heat exchanger, as shown in Fig. 5-24. The specific heat of the material before and after the reaction is 3.0 kJ/(kg · K). If the $UA$ of this counterflow regenerative heat exchanger is 2.1 kW/K and the flow rate is 1.2 kg/s, what is the temperature $t$ leaving the heat exchanger?

   Ans.: 57.9°C.

**5-7** A condenser having a $UA$ value of 480 kW/K condenses steam at a temperature of 40°C. The cooling water enters at 20°C with a flow rate of 160 kg/s. What is the outlet temperature of the cooling water? The specific heat of water is 4.19 kJ/(kg · K).

   Ans.: 30.2°C.

**5-8** A heat exchanger with one shell pass and two tube passes (Fig. 5-10) uses seawater at 15°C, $c_p$ = 3.8 kJ/(kg · K), to cool a flow rate of fresh water of 1.6 kg/s entering at 40°C. The specific heat of the fresh water is 4.19 kJ/(kg · K). If the $UA$ of the heat exchanger is 10 kW/K, what must the flow rate of seawater be in order to cool the fresh water to 22.5°C?

   Ans.: 7 kg/s.

**5-9** A double-pipe heat exchanger serves as an oil cooler with oil flowing in one direction through the inner tube and cooling water in the opposite direction through the annulus. The oil flow rate is 0.63 kg/s, the oil has a specific heat of 1.68 kJ/(kg · K), the water flow rate is 0.5 kg/s, and its specific heat is 4.19 kJ/(kg · K). In a test of a prototype, oil entering at 78°C was cooled to 54°C when the entering water temperature was 30°C. The possibility of increasing the area of the heat exchanger by increasing the length of the double pipe is to be considered. If the flow rates, fluid properties, and entering temperatures remain unchanged, what will the expected outlet temperature of the oil be if the area is increased by 20 percent?

   Ans.: 51.3°C.

**5-10** To ventilate a factory building, 5 kg/s of factory air at a temperature of 27°C is exhausted, and an identical flow rate of outdoor air at a temperature of −12°C is introduced to take its place. To recover some of the heat of the exhaust air, heat exchangers are placed in the exhaust and ventilation air ducts, as shown in Fig. 5-25, and 2 kg/s of water is

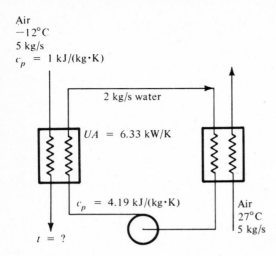

Air
−12°C
5 kg/s
$c_p = 1\ \text{kJ/(kg·K)}$

2 kg/s water

$UA = 6.33\ \text{kW/K}$

$c_p = 4.19\ \text{kJ/(kg·K)}$

$t = ?$

Air
27°C
5 kg/s

**Figure 5-25** Heat-recovery system in Prob. 5-10.

pumped between the two heat exchangers. The $UA$ value of both of these counterflow heat exchangers is 6.33 kW/K. What is the temperature of air entering the factory?
   **Ans.:** 2.9°C.

**5-11** A solar air heater consists of a flat air duct composed on one side of an absorbing sheet backed by insulation and on the other side by a transparent sheet, as shown in Fig. 5-26. The absorbing sheet absorbs 500 W/m² and delivers all this heat to the air being heated, which loses some to the atmosphere through the transparent sheet. The convection heat-transfer coefficient from the transparent sheet to the ambient air is 12 W/(m² · K) and from the air being heated to the transparent sheet is 45 W/(m² · K). The air enters with a temperature that is the same as the ambient, namely 15°C, and the flow rate of air is 0.02 kg/s per meter width. Develop the equation for the temperature of heated air $t$ as a function of length along the collector $x$ assuming no conduction in the sheet in the direction of airflow.
   **Ans.:** $t = 15 + 52.8(1 - e^{-x/2.11})$.

**5-12** The chain of heat exchangers shown in Fig. 5-27 has the purpose of elevating the temperature of a fluid to 390 K, at which temperature the desired chemical reaction takes place. The fluid has a specific heat of 3.2 kJ/(kg · K) both before and after the reaction, and

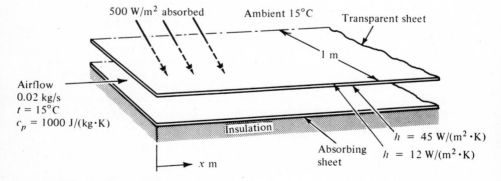

500 W/m² absorbed          Ambient 15°C          Transparent sheet

1 m

Airflow
0.02 kg/s
$t = 15°C$
$c_p = 1000\ \text{J/(kg·K)}$

Insulation

$x$ m

Absorbing sheet

$h = 45\ \text{W/(m²·K)}$
$h = 12\ \text{W/(m²·K)}$

**Figure 5-26** Solar air heater in Prob. 5-11.

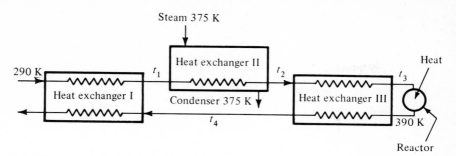

**Figure 5-27** Chain of heat exchangers in Prob. 5-12.

its flow rate is 1.5 kg/s. The entering temperature of the fluid to heat exchanger I is 290 K and the $UA$ of this heat exchanger is 2.88 kW/K. Steam is supplied to heat exchanger II at 375 K, and condensate leaves at the same temperature. The $UA$ values of heat exchangers II and III are 4.7 and 9.6 kW/K, respectively. What are the values of temperatures $t_1$ to $t_4$?

Ans.: $t_4$ = 365 K.

**5-13** A single-stage still, as in Fig. 5-18, is supplied with a feed of 0.6 mole fraction of $n$-butane and 0.4 mole fraction of $n$-heptane (Fig. 5-16). The still operates at a pressure of 700 kPa. How many moles of vapor are derived from 1 mol of feed if the vapor is to leave the still with a mole fraction of butane of 0.8?

Ans.: 0.57 mol.

**5-14** A vapor mixture of $n$-butane and $n$-heptane (Fig. 5-16) at a pressure of 700 kPa and a temperature of 170°C, and a mole fraction of butane of 0.4 enters a condenser.

(*a*) At what temperature does condensation begin?

(*b*) At what temperature is condensation complete?

(*c*) When the temperature is 120°C, what is the fraction in liquid form?

Ans.: (*c*) 0.76.

**5-15** A mixture of butane and propane is often sold as a fuel. We are interested in determining the $T$-vs.-$x$ relationship of a binary mixture of butane and propane at standard atmospheric pressure of 101.3 kPa. The pressure-temperature relationships at saturated conditions for the pure substances are

$$\ln P = \begin{cases} 21.40 - \dfrac{2286}{T} & \text{propane} \\[2mm] 21.77 - \dfrac{2795}{T} & \text{butane} \end{cases}$$

where $P$ = pressure, Pa

$T$ = absolute temperature, K

Present the $T$-vs.-$x$ curves for vapor and liquid neatly on a full-size sheet of graph paper, where $x$ represents the mole fraction of propane.

Ans.: One point on vapor curve, $x$ = 0.5 when $T$ = 260 K.

**5-16** A distillation tower (Fig. 5-28) receives a two-component solution in liquid form. The two components are designated as A and B, and $x$ indicates the mass fraction of material A. The concentration of the feed $x_1$ = 0.46 and the enthalpies entering and leaving the still are $h_1$ = 80 kJ/kg, $h_3$ = 360 kJ/kg, and $h_5$ = 97 kJ/kg. The condenser operates at $t$ = 30°C, at which state $x_2$ = 0.92, $h_2$ = 320 kJ/kg, $x_4$ = 0.82, $h_4$ = 23 kJ/kg, and the condenser rejects

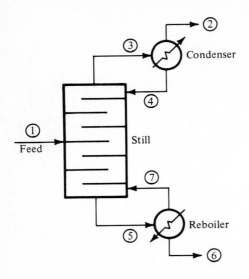

**Figure 5-28** Distillation tower in Prob. 5-16.

**Table 5-1 Operating conditions of distillation tower**

| Position | Flow rate, kg/s | $x$ | $h$, kJ/kg |
|----------|-----------------|------|------------|
| 1 | | 0.46 | 80 |
| 2 | | 0.92 | 320 |
| 3 | | | 360 |
| 4 | | 0.82 | 23 |
| 5 | | | 97 |
| 6 | | 0.08 | 108 |
| 7 | | 0.13 | 415 |

550 kW to the cooling water. The reboiler operates at $t = 210°C$, at which temperature $x_6 = 0.08$, $h_6 = 108$ kJ/kg, $x_7 = 0.13$, $h_7 = 415$ kJ/kg, and the reboiler receives 820 kW from high-pressure steam. Complete Table 5-1.

**Ans.:** Flow rate at 1 = 2.18 kg/s.

**5-17** Dimensional analysis suggests that the performance of a centrifugal fan can be expressed as a function of two dimensionless groups:

$$\frac{SP}{D^2 \omega^2 \rho} \quad \text{and} \quad \frac{Q}{D^3 \omega}$$

where SP = static pressure, Pa
    $D$ = diameter of wheel, m
    $\omega$ = rotative speed, rad/s
    $\rho$ = density, kg/m$^3$
    $Q$ = volume rate of airflow, m$^3$/s

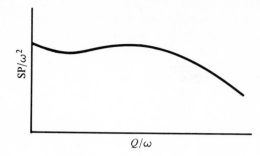

Figure 5-29 Performance of a centrifugal fan.

**Table 5-2  Performance of fan in Prob. 5-17**

| rad/s | $Q$, m$^3$/s | SP, Pa | rad/s | $Q$, m$^3$/s | SP, Pa |
|-------|--------------|--------|-------|--------------|--------|
| 157   | 1.42         | 861    | 126   | 3.30         | 114    |
|       | 1.89         | 861    | 94    | 0.94         | 304    |
|       | 2.36         | 796    |       | 1.27         | 299    |
|       | 2.83         | 694    |       | 1.89         | 219    |
|       | 3.02         | 635    |       | 2.22         | 134    |
|       | 3.30         | 525    |       | 2.36         | 100    |
| 126   | 1.42         | 548    | 63    | 0.80         | 134    |
|       | 1.79         | 530    |       | 1.04         | 122    |
|       | 2.17         | 473    |       | 1.42         | 70     |
|       | 2.36         | 428    |       | 1.51         | 55     |
|       | 2.60         | 351    |       |              |        |

For a given fan operating with air at a constant density, it should be possible to plot one curve, as in Fig. 5-29, that represents the performance at all speeds. The performance of a certain 0.3-m-diameter fan of Lau Blower Company is presented in Table 5-2.

(*a*) Plot neatly on graph paper the above performance data in the form of Fig. 5-29.

(*b*) If the SP is to be computed as a function of $Q$ and $\omega$, propose a convenient form of the equation (just use symbols for the coefficients; do not evaluate them numerically).

# REFERENCES

1. W. M. Kays and A. L. London, *Compact Heat Exchangers*, 2d ed., McGraw-Hill, New York, 1964.
2. P. Worsøe-Schmidt and H. J. Høgaard Knudsen, "Thermal Modelling of Heat Exchangers for Simulation Purposes," *25th Heat Transfer Fluid Mech. Inst., Davis, Calif., June 1976.*
3. W. Badger and J. T. Banchero, *Introduction to Chemical Engineering*, McGraw-Hill, New York, 1955.
4. W. L. McCabe and J. C. Smith, *Unit Operations of Chemical Engineering*, 3d ed., McGraw-Hill, New York, 1976.
5. C. M. Thatcher, *Fundamentals of Chemical Engineering*, Merrill, Columbus, Ohio, 1962.
6. C. G. Kirkbride, *Chemical Engineering Fundamentals*, McGraw-Hill, New York, 1947.

# SIX

# SYSTEM SIMULATION

## 6-1 DESCRIPTION OF SYSTEM SIMULATION

System simulation, as practiced in this chapter, is the calculation of operating variables (such as pressures, temperatures, and flow rates of energy and fluids) in a thermal system operating in a steady state. System simulation presumes knowledge of the performance characteristics of all components as well as equations for thermodynamic properties of the working substances. The equations for performance characteristics of the components and thermodynamic properties, along with energy and mass balances, form a set of simultaneous equations relating the operating variables. The mathematical description of system simulation is that of solving these simultaneous equations, many of which may be nonlinear.

A *system* is a collection of components whose performance parameters are interrelated. *System simulation* means observing a synthetic system that imitates the performance of a real system. The type of simulation studied in this chapter can be accomplished by calculation procedures, in contrast to simulating one physical system by observing the performance of another physical system. An example of two corresponding physical systems is when an electrical system of resistors and capacitors represents the heat-flow system in a solid wall.

## 6-2 SOME USES OF SIMULATION

System simulation may be used in the design stage to help achieve an improved design, or it may be applied to an existing system to explore prospective modifications. Simulation is not needed at the *design conditions* because in the design process the engineer probably chooses reasonable values of the operating variables (pressures, temperatures, flow rates, etc.) and selects the components (pumps, compressors, heat exchangers, etc.) that correspond to the operating variables. It would be for the nondesign conditions that system simulation would be applied, e.g., as at part-load or overload conditions. The designer may wish to investigate off-design operation to be sure that pressures, temperatures, or flow rates will not be too high or too low.

The steep increase in the cost of energy has probably been responsible for the blossoming of system simulation during recent years. Thermal systems (power generation, thermal processing, heating, and refrigeration) operate most of the time at off-design conditions. To perform energy studies in the design stage the operation of the system must be simulated throughout the range of operation the system will experience.

System simulation is sometimes applied to existing systems when there is an operating problem or a possible improvement is being considered. The effect on the system of changing a component can be examined before the actual change to ensure that the operating problem will be corrected and to find the cheapest means of achieving the desired improvement.

After listing some of the classes of system simulation, this chapter concentrates on just one class for the remainder of the chapter. Next the use of information-flow diagrams and the application to sequential and simultaneous calculations are discussed. The process of simulating thermal systems operating at steady state usually simmers down to the solution of simultaneous nonlinear algebraic equations, and procedures for their solution are examined.

## 6-3 CLASSES OF SIMULATION

System simulation is a popular term and is used in different senses by various workers. We shall first list some of the classes of system simulation and then designate the type to which our attention will be confined.

Systems may be classified as *continuous* or *discrete*. In a continuous system, the flow through the system is that of a continuum, e.g., a fluid or even solid particles, flowing at such rates relative to particle sizes that the stream can be considered as a continuum. In discrete systems, the flow is treated as a certain number of integers. The analysis of the flow of people through a supermarket involving the time spent at various shopping areas and the checkout counter is a discrete system. Another example of a discrete-system analysis is that performed in traffic control on expressways and city streets. Our concern, since it is primarily directed toward fluid and energy systems, is continuous systems.

Another classification is *deterministic* vs. *stochastic*. In the deterministic analysis the input variables are precisely specified. In stochastic analysis the input conditions are uncertain, either being completely random or (more commonly) following some probability distribution. In simulating the performance of a steam-electric generating plant that supplies both process steam and electric power to a facility, for example, a deterministic analysis starts with one specified value of the steam demand along with one specified value of the power demand. A stochastic analysis might begin with some probability description of the steam and power demands. We shall concentrate on deterministic analysis, although certainly a series of separate deterministic analyses could be made of different combinations of input conditions.

Finally, system simulation may be classed as *steady-state* or *dynamic*, where in a dynamic simulation there are changes of operating variables with respect to time. Dynamic analyses are used for such purposes as the study of a control system in order to achieve greater precision of control and to avoid unstable operating conditions. The dynamic simulation of a given system is more difficult than the steady-state simulation, since the steady state falls out as one special case of the transient analysis. On the other hand, steady-state simulations are required much more often than dynamic simulations and are normally applied to much larger systems.

The simulation to be practiced here will be that of continuous, deterministic steady-state systems.

## 6-4 INFORMATION-FLOW DIAGRAMS

Fluid- and energy-flow diagrams are standard engineering tools. In system simulation, an equally useful tool is the information-flow diagram. A block diagram of a control system is an information-flow diagram in which a block signifies that an output can be calculated when the input is known. In the block diagram used in automatic-control work the blocks represent *transfer functions*, which could be considered differential equations. In steady-state system simulation the block represents an algebraic equation. A centrifugal pump might appear in a fluid-flow diagram like that shown in Fig. 6-1$a$, while in the information-flow diagram the blocks (Fig. 6-1$b$) represent functions or expressions that permit

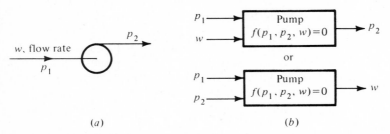

**Figure 6-1** ($a$) Centrifugal pump in fluid-flow diagram; ($b$) possible information-flow blocks representing pump.

calculation of the outlet pressure for the one block and the flow rate for the other. A block, as in Fig. 6-1b is usually an equation, here designated as $f(p_1, p_2, w) = 0$, or it may be tabular data to which interpolation would be applicable.

Figure 6-1 shows only one component. To illustrate how these individual blocks can build the information-flow diagram for a system, consider the fire-water facility shown in Fig. 6-2. A pump having pressure-flow characteristics shown in Fig. 6-2 draws water from an open reservoir and delivers it through a length of pipe to hydrant $A$, with some water continuing through additional pipe to hydrant $B$. The water flow rates in the pipe sections are designated $w_1$ and $w_2$, and the flow rates passing out the hydrants are $w_A$ and $w_B$. The equations for the water flow rate through open hydrants are $w_A = C_A\sqrt{p_3 - p_{at}}$ and $w_B = C_B\sqrt{p_4 - p_{at}}$, where $C_A$ and $C_B$ are constants and $p_{at}$ is the atmospheric pressure. The equation for the pipe section 0-1 is $p_{at} - p_1 = C_1 w_1^2 + h\rho g$, where $C_1 w_1^2$ accounts for friction and $h\rho g$ is the pressure drop due to the change in elevation $h$. In pipe sections 2-3 and 3-4

$$p_2 - p_3 = C_2 w_1^2 \quad \text{and} \quad p_3 - p_4 = C_3 w_2^2$$

These five equations can be written in functional form

$$f_1(w_A, p_3) = 0 \tag{6-1}$$

$$f_2(w_B, p_4) = 0 \tag{6-2}$$

$$f_3(w_1, p_1) = 0 \tag{6-3}$$

$$f_4(w_1, p_2, p_3) = 0 \tag{6-4}$$

$$f_5(w_2, p_3, p_4) = 0 \tag{6-5}$$

The atmospheric pressure $p_{at}$ is not listed as a variable since it will have a known value. An additional function is provided by the pump characteristics

$$f_6(w_1, p_1, p_2) = 0 \tag{6-6}$$

The preceding six equations can all be designated as *component performance characteristics*. There are eight unknown variables, $w_1, w_2, w_A, w_B, p_1, p_2, p_3,$

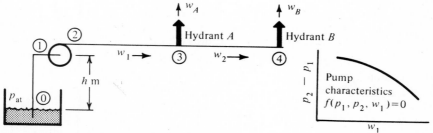

**Figure 6-2** Fire-water system and pump characteristics.

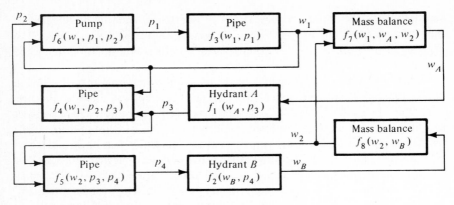

Figure 6-3 Information-flow diagram for fire-water system.

and $p_4$, but only six equations so far. Mass balances provide the other two equations

$$w_1 = w_A + w_2 \qquad \text{or} \qquad f_7(w_1, w_A, w_2) = 0 \qquad (6\text{-}7)$$

and

$$w_2 = w_B \qquad \text{or} \qquad f_8(w_2, w_B) = 0 \qquad (6\text{-}8)$$

Several correct flow diagrams can be developed to express this system, one of which is shown in Fig. 6-3. Each block is arranged so that there is only one output, which indicates that the equation represented by that block is solved for the output variable.

## 6-5 SEQUENTIAL AND SIMULTANEOUS CALCULATIONS

Sometimes it is possible to start with the input information and immediately calculate the output of a component. The output information from this first component is all that is needed to calculate the output information of the next component, and so on to the final component of the system, whose output is the output information of the system. Such a system simulation consists of *sequential calculations*. An example of a sequential calculation might occur in an on-site power-generating plant using heat recovery to generate steam for heating or refrigeration, as shown schematically in Fig. 6-4. The exhaust gas from the engine flows through the boiler, which generates steam to operate an absorption refrigeration unit. If the output information is the refrigeration capacity that would be available when the unit generates a given electric-power requirement, a possible information-flow diagram for this simulation is shown in Fig. 6-5.

Starting with the knowledge of the engine-generator speed and electric-power demand, we can solve the equations representing performance characteristics of the components in sequence to arrive at the output information, the refrigeration capacity.

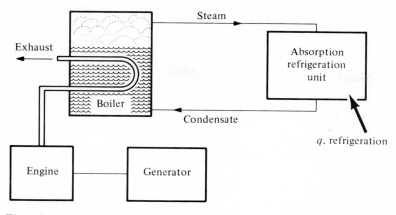

**Figure 6-4** On-site power generation with heat recovery to develop steam for refrigeration.

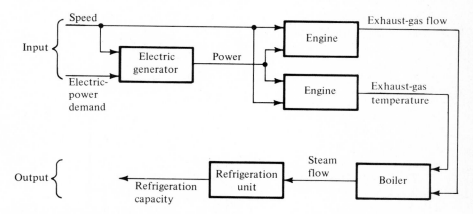

**Figure 6-5** Information-flow diagram for on-site power-generating plant of Fig. 6-4.

The *sequential* simulation shown by the information-flow diagram of Fig. 6-5 is in contrast to the *simultaneous* simulation required for the information-flow diagram of Fig. 6-3. Sequential simulations are straightforward, but simultaneous simulations are the challenges on which the remainder of the chapter concentrates.

## 6-6 TWO METHODS OF SIMULATION: SUCCESSIVE SUBSTITUTION AND NEWTON-RAPHSON

The task of simulating a system, after the functional relationships and interconnections have been established, is one of solving a set of simultaneous algebraic equations, some or all of which may be nonlinear. Two of the methods available for this simultaneous solution are *successive substitution* and *Newton-Raphson*. Each method has its advantages and disadvantages which will be pointed out.

## 6-7 SUCCESSIVE SUBSTITUTION

The method of successive substitution is closely associated with the information-flow diagram of the system (Fig. 6-3). There seems to be no way to find a toe-hold to begin the calculations. The problem is circumvented by assuming a value of one or more variables, beginning the calculation, and proceeding through the system until the originally-assumed variables have been recalculated. The recalculated values are substituted successively (which is the basis for the name of the method), and the calculation loop is repeated until satisfactory convergence is achieved.

**Example 6-1** A water-pumping system consists of two parallel pumps drawing water from a lower reservoir and delivering it to another that is 40 m higher, as illustrated in Fig. 6-6. In addition to overcoming the pressure difference due to the elevation, the friction in the pipe is $7.2w^2$ kPa, where $w$ is the combined flow rate in kilograms per second. The pressure–flow-rate characteristics of the pumps are

Pump 1: $\qquad\qquad \Delta p, \text{kPa} = 810 - 25w_1 - 3.75w_1^2$

Pump 2: $\qquad\qquad \Delta p, \text{kPa} = 900 - 65w_2 - 30w_2^2$

where $w_1$ and $w_2$ are the flow rates through pump 1 and pump 2, respectively.

Use successive substitution to simulate this system and determine the values of $\Delta p, w_1, w_2$, and $w$.

SOLUTION: The system can be represented by four simultaneous equations. The pressure difference due to elevation and friction is

$$\Delta p = 7.2w^2 + \frac{(40 \text{ m})(1000 \text{ kg/m}^3)(9.807 \text{ m/s}^2)}{1000 \text{ Pa/kPa}} \qquad (6\text{-}9)$$

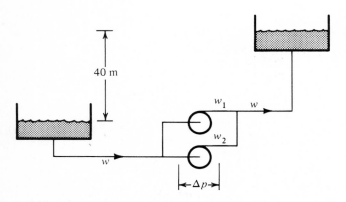

Figure 6-6 Water-pumping system in Example 6-1.

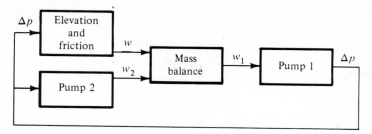

Figure 6-7 Information-flow diagram 1 for Example 6-1.

**Table 6-1 Successive substitution on information-flow diagram of Fig. 6-7**

| Iteration | $\Delta p$ | $w_2$ | $w$ | $w_1$ |
|---|---|---|---|---|
| 1 | 638.85 | 2.060 | 5.852 | 3.792 |
| 2 | 661.26 | 1.939 | 6.112 | 4.174 |
| 3 | 640.34 | 2.052 | 5.870 | 3.818 |
| 4 | 659.90 | 1.946 | 6.097 | 4.151 |
| . . . . . . . . . . . . . . . . . . . . . . . . . . . . . . . . . . . . . . . . |
| 47 | 649.98 | 2.000 | 5.983 | 3.983 |
| 48 | 650.96 | 1.995 | 5.994 | 3.999 |
| 49 | 650.04 | 2.000 | 5.983 | 3.984 |
| 50 | 650.90 | 1.995 | 5.993 | 3.998 |

Pump 1: $\qquad$ $\Delta p = 810 - 25w_1 - 3.75w_1^2$ $\qquad$ (6-10)

Pump 2: $\qquad$ $\Delta p = 900 - 65w_2 - 30w_2^2$ $\qquad$ (6-11)

Mass balance: $\qquad$ $w = w_1 + w_2$ $\qquad$ (6-12)

One possible information-flow diagram that represents this system is shown in Fig. 6-7. If a trial value of 4.2 is chosen for $w_1$, the value of $\Delta p$ can be computed from Eq. (6-10), and so on about the loop. The values of the variables resulting from these iterations are shown in Table 6-1. The calculation appears to be converging slowly to the values $w_1 = 3.991$, $w_2 = 1.99$, $w = 5.988$, and $\Delta p = 650.5$.

## 6-8 PITFALLS IN THE METHOD OF SUCCESSIVE SUBSTITUTION

Figure 6-7 is only one of the possible information-flow diagrams that can be generated from the set of equations (6-9) to (6-12). Two additional flow diagrams are shown in Figs. 6-8 and 6-9.

The trial value of $w_2 = 2.0$ was chosen for the successive substitution method on information-flow diagram 2, and the results of the iterations are shown in

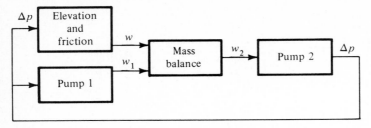

**Figure 6-8** Information-flow diagram 2 for Example 6-1.

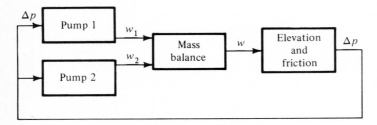

**Figure 6-9** Information-flow diagram 3 for Example 6-1.

Table 6-2. A trial value of $w = 6.0$ was chosen for the solution of information-flow diagram 3, and the results are shown in Table 6-3.

Information-flow diagram 1 converged to the solution, while diagrams 2 and 3 diverged. This experience is typical of successive substitution. It should be observed that the divergence in diagrams 2 and 3 is attributable to the calculation sequence and not a faulty choice of the trial value. In both cases the trial value was essentially the correct solution: $w_2 = 2.0$ and $w = 6.0$.

Are there means of checking a flow diagram in advance to determine whether the calculations will converge or diverge? Yes, but the effort of such a check is probably greater than simply experimenting with various diagrams until one is found that converges.

In the method of successive substitution each equation is solved for one

**Table 6-2  Iterations of information-flow diagram 2**

| Iteration | $\Delta p$ | $w$ | $w_1$ | $w_2$ |
|---|---|---|---|---|
| 1 | 650.0 | 5.983 | 4.000 | 1.983 |
| 2 | 653.2 | 6.019 | 3.942 | 2.077 |
| 3 | 635.5 | 5.812 | 4.258 | 1.554 |
| 4 | 726.5 | 6.814 | 2.443 | 4.371 |
| 5 | 42.8 | † | | |

†Value of $w$ became imaginary.

**Table 6-3 Iterations of information-flow diagram 3**

| Iteration | $\Delta p$ | $w_1$ | $w_2$ | $w$ |
|---|---|---|---|---|
| 1 | 651.5 | 3.973 | 1.992 | 5.965 |
| 2 | 648.5 | 4.028 | 2.008 | 6.036 |
| 3 | 654.6 | 3.916 | 1.975 | 5.891 |
| 4 | 642.1 | 4.142 | 2.042 | 6.184 |
| 5 | 667.6 | 3.672 | 1.903 | 5.575 |
| 6 | 616.1 | 4.593 | 2.178 | 6.771 |
| 7 | 722.3 | 2.539 | 1.580 | 4.120 |
| 8 | 514.5 | 6.149 | 2.662 | 8.811 |
| 9 | 951.2 | † | | |

†Value of $w_1$ became imaginary.

variable, and the equation may be nonlinear in that variable, as was true, for example, for the calculations of $w$ and $w_2$ in diagram 1. No particular problem resulted in computing $w$ and $w_2$ here because the equations were quadratic. An iterative technique, which may be required in some cases, is described in Sec. 6-10.

## 6-9 TAYLOR-SERIES EXPANSION

The second technique of system simulation, presented in Secs. 6-10 and 6-11, the Newton-Raphson method, is based on a Taylor-series expansion. It is therefore appropriate to review the Taylor-series expansion. If a function $z$, which is dependent upon two variables $x$ and $y$, is to be expanded about the point $(x = a, y = b)$, the form of the series expansion is

$$z = \text{const} + \text{first-degree terms} + \text{second-degree terms} + \text{higher-degree terms}$$

or, more specifically,

$$z = c_0 + [c_1(x - a) + c_2(y - b)] + [c_3(x - a)^2 + c_4(x - a)(y - b)$$
$$+ c_5(y - b)^2] + \cdots \quad (6\text{-}13)$$

Now determine the values of the constants in Eq. (6-13). If $x$ is set equal to $a$ and $y$ is set equal to $b$, all the terms on the right side of the equation reduce to zero except $c_0$, so that the value of the function at $(a, b)$ is

$$c_0 = z(a, b) \quad (6\text{-}14)$$

To find $c_1$, partially differentiate Eq. (6-13) with respect to $x$; then set $x = a$ and $y = b$. The only term remaining on the right side of Eq. (6-13) is $c_1$, so

$$c_1 = \frac{\partial z(a, b)}{\partial x} \quad (6\text{-}15)$$

In a similar manner,

$$c_2 = \frac{\partial z(a, b)}{\partial y} \tag{6-16}$$

The constants $c_3$, $c_4$, and $c_5$ are found by partial differentiation twice followed by substitution of $x = a$ and $y = b$ to yield

$$c_3 = \frac{1}{2} \frac{\partial^2 z(a, b)}{\partial x^2} \qquad c_4 = \frac{\partial^2 z(a, b)}{\partial x \, \partial y} \qquad c_5 = \frac{1}{2} \frac{\partial^2 z(a, b)}{\partial y^2} \tag{6-17}$$

For the special case where $y$ is a function of one independent variable $x$, the Taylor-series expansion about the point $x = a$ is

$$y = y(a) + \frac{dy(a)}{dx} (x - a) + \left[ \frac{1}{2} \frac{d^2 y(a)}{dx^2} \right] (x - a)^2 + \cdots \tag{6-18}$$

The general expression for the Taylor-series expansion if $y$ is a function of $n$ variables $x_1, x_2, \ldots, x_n$ around the point $(x_1 = a_1, x_2 = a_2, \ldots, x_n = a_n)$ is

$$y(x_1, x_2, \ldots, x_n) = y(a_1, a_2, \ldots, a_n) + \sum_{j=1}^{n} \frac{\partial y(a_1, \ldots, a_n)}{\partial x_j} (x_j - a_j)$$

$$+ \frac{1}{2} \sum_{i=1}^{n} \sum_{j=1}^{n} \frac{\partial^2 y(a_1, \ldots, a_n)}{\partial x_i \, \partial x_j} (x_i - a_i)(x_j - a_j) + \cdots \tag{6-19}$$

**Example 6-2** Express the function $\ln (x^2/y)$ as a Taylor-series expansion about the point $(x = 2, y = 1)$.

SOLUTION

$$z = \ln \frac{x^2}{y} = c_0 + c_1 (x - 2) + c_2 (y - 1) + c_3 (x - 2)^2 + c_4 (x - 2)(y - 1)$$

$$+ c_5 (y - 1)^2 + \cdots$$

Evaluating the constants, we get

$$c_0 = \ln \frac{2^2}{1} = \ln 4 = 1.39$$

$$c_1 = \frac{\partial z(2, 1)}{\partial x} = \frac{2x/y}{x^2/y} = \frac{2}{x} = 1$$

$$c_2 = \frac{\partial z(2, 1)}{\partial y} = -\frac{x^2/y^2}{x^2/y} = -\frac{1}{y} = -1$$

$$c_3 = \frac{1}{2} \frac{\partial^2 z(2, 1)}{\partial x^2} = \frac{1}{2} \left( -\frac{2}{x^2} \right) = -\frac{1}{4}$$

$$c_4 = \frac{\partial^2 z(2, 1)}{\partial x\, \partial y} = 0$$

$$c_5 = \frac{1}{2} \frac{\partial^2 z(2, 1)}{\partial y^2} = \frac{1}{2} \frac{1}{y^2} = \frac{1}{2}$$

The first several terms of the expansion are

$$z = 1.39 + (x - 2) - (y - 1) - (\tfrac{1}{4})(x - 2)^2 + (\tfrac{1}{2})(y - 1)^2 + \cdots$$

## 6-10 NEWTON-RAPHSON WITH ONE EQUATION AND ONE UNKNOWN

In the Taylor-series expansion of Eq. (6-18) when $x$ is close to $a$, the higher-order terms become negligible. The equation then reduces approximately to

$$y \approx y(a) + [y'(a)](x - a) \qquad (6\text{-}20)$$

Equation (6-20) is the basis of the Newton-Raphson iterative technique for solving a nonlinear algebraic equation. Suppose that the value of $x$ is sought that satisfies the equation

$$x + 2 = e^x \qquad (6\text{-}21)$$

Define $y$ as

$$y(x) = x + 2 - e^x \qquad (6\text{-}22)$$

and denote $x_c$ as the correct value of $x$ that solves Eq. (6-21) and makes $y = 0$

$$y(x_c) = 0 \qquad (6\text{-}23)$$

The Newton-Raphson process requires an initial assumption of the value of $x$. Denote as $x_t$ this temporary value of $x$. Substituting $x_t$ into Eq. (6-22) gives a value of $y$ which almost certainly does not provide the desired value of $y = 0$. Specifically, if $x_t = 2$,

$$y(x_t) = x_t + 2 - e^{x_t} = 2 + 2 - 7.39 = -3.39$$

Our trial value of $x$ is incorrect, but now the question is how the value of $x$ should be changed in order to bring $y$ closer to zero.

Returning to the Taylor expansion of Eq. (6-20), express $y$ in terms of $x$ by expanding about $x_c$

$$y(x) \approx y(x_c) + [y'(x_c)](x - x_c) \qquad (6\text{-}24)$$

For $x = x_t$, Eq. (6-24) becomes

$$y(x_t) \approx y(x_c) + [y'(x_t)](x_t - x_c) \qquad (6\text{-}25)$$

Equation (6-25) contains the further approximation of evaluating the derivative at $x_t$ rather than at $x_c$, because the value of $x_c$ is still unknown. From Eq. (6-23)

$y(x_c) = 0$, and so Eq. (6-25) can be solved approximately for the unknown value of $x_c$

$$x_c \approx x_t - \frac{y(x_t)}{y'(x_t)} \tag{6-26}$$

In the numerical example

$$x_c \approx 2 - \frac{-3.39}{1 - e^2} = 1.469$$

The value of $x = 1.469$ is a more correct value and should be used for the next iteration. The results of the next several iterations are

| $x$ | 1.469 | 1.208 | 1.152 |
|---|---|---|---|
| $y(x)$ | −0.876 | −0.132 | −0.018 |

The graphic visualization of the iteration is shown in Fig. 6-10, where we seek the root of the equation $y = x + 2 - e^x$. The first trial is at $x = 2$, and the deviation of $y$ from zero divided by the slope of the curve there suggests a new trial of 1.469.

The Newton-Raphson method, while it is a powerful iteration technique, should be used carefully because if the initial trial is too far off from the correct result, the solution may not converge. Some insight into the nature of the function being solved is therefore always helpful.

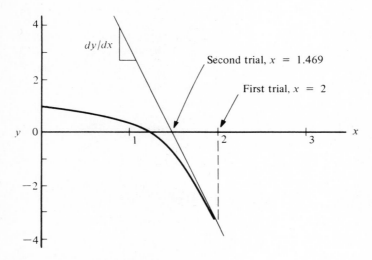

**Figure 6-10** Newton-Raphson iteration.

## 6-11 NEWTON-RAPHSON METHOD WITH MULTIPLE EQUATIONS AND UNKNOWNS

The solution of a nonlinear equation for the unknown variable discussed in Sec. 6-10 is only a special case of the solution of a set of multiple nonlinear equations. Suppose that three nonlinear equations are to be solved for the three unknown variables $x_1, x_2$, and $x_3$

$$f_1(x_1, x_2, x_3) = 0 \qquad (6\text{-}27)$$

$$f_2(x_1, x_2, x_3) = 0 \qquad (6\text{-}28)$$

$$f_3(x_1, x_2, x_3) = 0 \qquad (6\text{-}29)$$

The procedure for solving the equations is an iterative one in which the following steps are followed:

1. Rewrite the equations so that all terms are on one side of the equality sign [Eqs. (6-27) to (6-29) already exist in this form].
2. Assume temporary values for the variables, denoted $x_{1,t}, x_{2,t}$, and $x_{3,t}$.
3. Calculate the values of $f_1, f_2$, and $f_3$ at the temporary values of $x_1, x_2$, and $x_3$.
4. Compute the partial derivatives of all functions with respect to all variables.
5. Use the Taylor-series expansion of the form of Eq. (6-19) to establish a set of simultaneous equations. The Taylor-series expansion for Eq. (6-27), for example, is

$$
\begin{aligned}
f_1(x_{1,t}, x_{2,t}, x_{3,t}) \approx{} & f_1(x_{1,c}, x_{2,c}, x_{3,c}) \\
& + \frac{\partial f_1(x_{1,t}, x_{2,t}, x_{3,t})}{\partial x_1}(x_{1,t} - x_{1,c}) \\
& + \frac{\partial f_1(x_{1,t}, x_{2,t}, x_{3,t})}{\partial x_2}(x_{2,t} - x_{2,c}) \\
& + \frac{\partial f_1(x_{1,t}, x_{2,t}, x_{3,t})}{\partial x_3}(x_{3,t} - x_{3,c}) \qquad (6\text{-}30)
\end{aligned}
$$

The set of equations is

$$
\begin{bmatrix}
\dfrac{\partial f_1}{\partial x_1} & \dfrac{\partial f_1}{\partial x_2} & \dfrac{\partial f_1}{\partial x_3} \\[2mm]
\dfrac{\partial f_2}{\partial x_1} & \dfrac{\partial f_2}{\partial x_2} & \dfrac{\partial f_2}{\partial x_3} \\[2mm]
\dfrac{\partial f_3}{\partial x_1} & \dfrac{\partial f_3}{\partial x_2} & \dfrac{\partial f_3}{\partial x_3}
\end{bmatrix}
\begin{bmatrix}
x_{1,t} - x_{1,c} \\[2mm]
x_{2,t} - x_{2,c} \\[2mm]
x_{3,t} - x_{3,c}
\end{bmatrix}
=
\begin{bmatrix}
f_1 \\[2mm]
f_2 \\[2mm]
f_3
\end{bmatrix}
\qquad (6\text{-}31)
$$

6. Solve the set of linear simultaneous equations (6-31) to determine $x_{i,t} - x_{i,c}$.

7. Correct the $x$'s

$$x_{1,new} = x_{1,old} - (x_{1,t} - x_{1,c})$$

$$\cdots\cdots\cdots\cdots\cdots\cdots\cdots\cdots$$

$$x_{3,new} = x_{3,old} - (x_{3,t} - x_{3,c})$$

8. Test for convergence. If the absolute magnitudes of all the $f$'s or all the $\Delta x$'s are satisfactorily small, terminate; otherwise return to step 3.

**Example 6-3** Solve Example 6-1 by the Newton-Raphson method.

SOLUTION

Step 1.
$$f_1 = \Delta p - 7.2w^2 - 392.28 = 0$$

$$f_2 = \Delta p - 810 + 25w_1 + 3.75w_1^2 = 0$$

$$f_3 = \Delta p - 900 + 65w_2 + 30w_2^2 = 0$$

$$f_4 = w - w_1 - w_2 = 0$$

Step 2. Choose trial values of the variables, which are here selected as $\Delta p = 750$, $w_1 = 3$, $w_2 = 1.5$, and $w = 5$.

Step 3. Calculate the magnitudes of the $f$'s at the temporary values of the variables, $f_1 = 177.7$, $f_2 = 48.75$, $f_3 = 15.0$, and $f_4 = 0.50$.

Step 4. The partial derivatives are shown in Table 6-4.

Step 5. Substituting the temporary values of the variables into the equations for the partial derivatives forms a set of linear simultaneous equations to be solved for the corrections to $x$:

$$\begin{bmatrix} 1.0 & 0.0 & 0.0 & -72.0 \\ 1.0 & 47.5 & 0.0 & 0.0 \\ 1.0 & 0.0 & 155.0 & 0.0 \\ 0.0 & -1.0 & -1.0 & 1.0 \end{bmatrix} \begin{bmatrix} \Delta x_1 \\ \Delta x_2 \\ \Delta x_3 \\ \Delta x_4 \end{bmatrix} = \begin{bmatrix} 177.7 \\ 48.75 \\ 15.0 \\ 0.50 \end{bmatrix}$$

where
$$\Delta x_i = x_{i,t} - x_{i,c}$$

Step 6. Solution of the simultaneous equations

$$\Delta x_1 = 98.84 \qquad \Delta x_2 = -1.055 \qquad \Delta x_3 = -0.541 \qquad \Delta x_4 = -1.096$$

**Table 6-4**

| | $\partial /\partial\Delta p$ | $\partial /\partial w_1$ | $\partial /\partial w_2$ | $\partial /\partial w$ |
|---|---|---|---|---|
| $\partial f_1/\partial$ | 1 | 0 | 0 | $-14.4w$ |
| $\partial f_2/\partial$ | 1 | $25 + 7.5w_1$ | 0 | 0 |
| $\partial f_3/\partial$ | 1 | 0 | $65 + 60w_2$ | 0 |
| $\partial f_4/\partial$ | 0 | $-1$ | $-1$ | 1 |

**Table 6-5**

| After iteration | $f_1$ | $f_2$ | $f_3$ | $f_4$ | $\Delta p$ | $w_1$ | $w_2$ | $w$ |
|---|---|---|---|---|---|---|---|---|
| 1 | −8.641 | 4.170 | 8.778 | 0.000 | 651.16 | 4.055 | 2.041 | 6.096 |
| 2 | −0.081 | 0.0148 | 0.056 | 0.000 | 650.48 | 3.992 | 1.998 | 5.989 |
| 3 | 0.000 | 0.000 | 0.000 | 0.000 | 650.49 | 3.991 | 1.997 | 5.988 |

*Step 7.* The corrected values of the variables are

$$\Delta p = 750.0 - 98.84 = 651.16 \quad w_1 = 4.055 \quad w_2 = 2.041 \quad w = 6.096$$

These values of the variables are returned to step 3 for the next iteration.

The values of the $f$'s and the variables resulting from continued iterations are shown in Table 6-5.

The calculations converged satisfactorily after three iterations.

## 6-12 SIMULATION OF A GAS TURBINE SYSTEM

A simulation of a more extensive thermal system will be given for a nonregenerative gas-turbine cycle. This cycle, shown in Fig. 6-11, consists of a compressor, combustor, and turbine whose performance characteristics are known. The turbine-compressor combination operates at 120 r/s.

The objective of the simulation is to determine the power output at the shaft, $E_s$ kW, if 8000 kW of energy is added at the combustor by burning fuel. The turbine draws air and rejects the turbine exhaust to atmospheric pressure of 101 kPa. The entering air temperature is 25°C. Certain simplifications will be

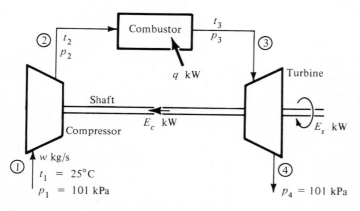

**Figure 6-11** Gas-turbine cycle.

introduced in the solution, but it is understood that the simulation method can be extended to more refined calculations. The simplifications are

1. Assume perfect-gas properties throughout the cycle and a $c_p$ constant at 1.03 kJ/(kg · K)
2. Neglect the mass added in the form of fuel in the combustor so that the mass rate of flow $w$ is constant throughout the cycle
3. Neglect the pressure drop in the combustor so that $p_2 = p_3$ and the high pressure in the system can be designated simply as $p$
4. Neglect heat transfer to the environment

The performance characteristics of the axial-flow compressor and the gas turbine[1] operating at 120 r/s with an atmospheric pressure of 101 kPa that will be used in the simulation are shown in Figs. 6-12 and 6-13, respectively.

With the techniques presented in Chap. 4 equations can be developed for the curves in Fig. 6-12,

$$p = 331 + 45.6w - 4.03w^2 \tag{6-32}$$

and
$$E_c = 1020 - 0.383p + 0.00513p^2 \tag{6-33}$$

where $p$ = discharge pressure of compressor, kPa
$\quad\ w$ = mass rate of flow, kg/s
$\quad\ E_c$ = power required by compressor, kW

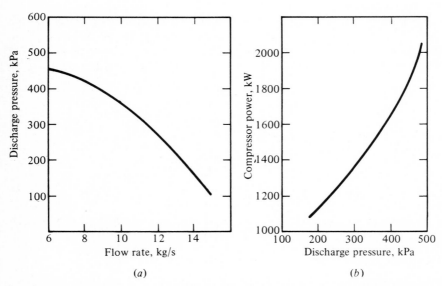

Figure 6-12 Performance of axial-flow compressor operating at 120 r/s with 101 kPa inlet pressure.

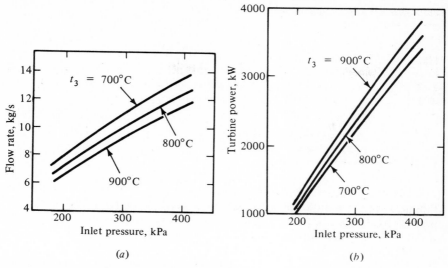

**Figure 6-13** Performance of gas turbine operating at 120 r/s and 101 kPa discharge pressure.

When operating at a given speed and discharge pressure, the characteristics of the turbine take the form shown in Fig. 6-13. With the techniques of Sec. 4-8 equations can be developed for the curves in Fig. 6-13

$$w = 8.5019 + 0.02332p + 0.48 \times 10^{-4}p^2 - 0.02644t$$
$$+ 0.1849 \times 10^{-4}t^2 + 0.000121pt - 0.2736 \times 10^{-6}p^2 t$$
$$- 0.1137 \times 10^{-6}pt^2 + 0.2124 \times 10^{-9}p^2 t^2 \tag{6-34}$$

and
$$E_t = 1727.5 - 10.06p + 0.033033p^2 - 7.4709t + 0.003919t^2$$
$$+ 0.050921pt - 0.8525 \times 10^{-4}p^2 t - 0.2356 \times 10^{-4}pt^2$$
$$+ 0.4473 \times 10^{-7}p^2 t^2 \tag{6-35}$$

where $t$ = entering temperature = $t_3$, °C
   $E_t$ = power delivered by turbine, kW

To achieve the simulation the values of the following unknown variables must be determined: $w$, $p$, $E_c$, $t_2$, $E_s$, $t_3$, and $E_t$. Seven independent equations must be found to solve for this set of unknowns. Four equations are available from the performance characteristics of the compressor and turbine. The three other equations come from energy balances:

Compressor: $\qquad\qquad\qquad E_c = wc_p(t_2 - 25) \qquad\qquad$ (6-36)

Combustor: $\qquad\qquad\qquad 8000 = wc_p(t_3 - t_2) \qquad\qquad$ (6-37)

Turbine power: $\qquad\qquad\qquad E_t = E_c + E_s \qquad\qquad$ (6-38)

**Table 6-6 Newton-Raphson solution of gas-turbine simulation**

|  | $w$ | $p$ | $E_c$ | $t_2$ | $E_s$ | $t_3$ | $E_t$ |
|---|---|---|---|---|---|---|---|
| Trial value | 10.00 | 300.0 | 1000.0 | 250.0 | 1500.0 | 800.0 | 2400.0 |
| After iteration: |  |  |  |  |  |  |  |
| 1 | 10.91 | 352.2 | 1507.8 | 150.8 | 1569.3 | 877.5 | 3077.0 |
| 2 | 10.77 | 354.8 | 1530.0 | 162.8 | 1597.7 | 884.2 | 3127.7 |
| 3 | 10.77 | 354.9 | 1530.1 | 173.0 | 1598.5 | 884.5 | 3128.6 |
| 4 | 10.77 | 354.9 | 1530.1 | 173.0 | 1598.5 | 884.5 | 3128.6 |

The execution of the solution follows the steps outlined in Sec. 6-11. A summary of the trial values and results after the Newton-Raphson iterations is presented in Table 6-6.

The shaft power delivered by this system is 1598.5 kW.

## 6-13 OVERVIEW OF SYSTEM SIMULATION

Steady-state simulation of thermal systems is fast increasing in applicability. The uses of simulation include evaluation of part-load operation directed toward identifying potential operating problems and also predicting annual energy requirements of systems.[2] System simulation can also be one step in an optimization process. For example, the effect on the output of the system of making a small change in one component, e.g., the size of a heat exchanger, is essentially a partial derivative of the type that will be needed in certain of the optimization techniques to be explained in later chapters.

If the exposure to system simulation in this chapter was the reader's first experience with it, wrestling with the techniques may be the major preoccupation. After the methods have been mastered, setting up the equations becomes the major challenge. In large systems it may not be simple to choose the proper combination of equations that precisely specifies the system while avoiding combinations of dependent equations. Unfortunately, no methodical procedure has yet been developed for choosing the equations; a thorough grounding in thermal principles and a bit of intuition are still the necessary tools.

The mathematical description of steady-state system simulation is that of solving a simultaneous set of algebraic equations, some of which are nonlinear. One impulse might be first to eliminate equations and variables by substitution. This strategy is not normally recommended when using a computer to perform the successive substitution or Newton-Raphson solution. Working with the full set of equations provides the solution to a larger number of variables directly, some of which may be of interest. Performing the substitution always presents the hazard of making an algebraic error, and the equations in combined form are more difficult to check than their simpler basic arrangement.

Two methods of system simulation have been presented in this chapter, successive substitution and Newton-Raphson. Successive substitution is a straightforward technique and is usually easy to program. It uses computer memory sparingly. The disadvantages are that sometimes the sequence may either converge very slowly or diverge. As far as can be determined through the web of commercial secrecy, many of the large simulation programs used in the petroleum, chemical, and thermal processing industries rely heavily on the successive-substitution method.[3,4] The experienced programmer will enhance his chances of a convergent sequence by choosing the blocks in the information-flow diagram in such a way that the output is only moderately affected by large changes in the input. A fruitful area for continued study is to develop methods of accelerating convergence or damping divergence.

The Newton-Raphson technique, while a bit more complex, is quite powerful. The drudgery of computing the partial derivatives can be circumvented by allowing the computer to extract the partial derivatives numerically.[5] A challenge that arises in simulating large systems is that the number of equations becomes so large that the $n \times n$ matrix (where $n$ is the number of equations and unknowns) needed for solving the equation analogous to Eq. (6-31) requires excessive memory space in the computer. Techniques for solving linear simultaneous equations that have sparse matrices are called for. Another approach is to break the system apart into smaller subsystems, which can be simulated individually and their results tied together by successive substitution.

## PROBLEMS

**6-1** The operating point of a fan-and-duct system is to be determined. The equations for the two components are

Duct: $\qquad SP = 80 + 10.73Q^{1.8}$

Fan: $\qquad Q = 15 - (73.5 \times 10^{-6})SP^2$

where SP = static pressure, Pa
$\quad Q$ = airflow rate, $m^3/s$

Use successive substitution to solve for the operating point, choosing as trial values SP = 200 Pa or $Q = 10 \ m^3/s$.
$\qquad$ Ans.: 6 $m^3/s$ and 350 Pa.

**6-2** The torque–rotative-speed curve of the engine-and-drive train of a truck operating at a certain transmission setting is shown in Fig. 6-14. The $T$ vs. $\omega$ curve for the load on the truck is also shown and is appropriate for the truck moving slowly uphill. The equations for the two curves are

Engine drive: $\qquad T = -170 + 29.4\omega - 0.284\omega^2$

Load: $\qquad T = 10.5\omega$

where $T$ = torque, N · m
$\quad \omega$ = rotative speed, r/s

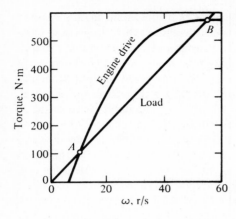

**Figure 6-14** Torque–rotative-speed curves of engine drive and load on a truck.

**Table 6-7 Results of iterations in Prob. 6-2**

| Flow diagram Fig. 6-15a | | | Flow diagram Fig. 6-15b | | |
|---|---|---|---|---|---|
| Iteration | $T$ | $\omega$ | Iteration | $T$ | $\omega$ |
| 0 | ✕ | 40 | 0 | ✕ | 40 |
| 1 | | | 1 | | |
| 2 | | | 2 | | |
| 3 | | | 3 | | |

☐ Converging to point ___     ☐ Converging to point ___
☐ Diverging                            ☐ Diverging

(*a*) Determination of the operating condition of the truck is a simulation of a two-component system. Perform this simulation with both flow diagrams shown in Fig. 6-15. Use an initial value for both simulations of $\omega = 40$ r/s and show the results in the form of Table 6-7.

(*b*) From a physical standpoint, explain the behavior of the system when operating in the immediate vicinity (on either side) of $A$.

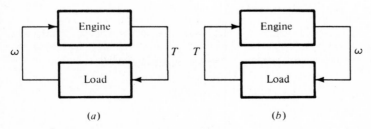

**Figure 6-15** Flow diagrams in Prob. 6-2.

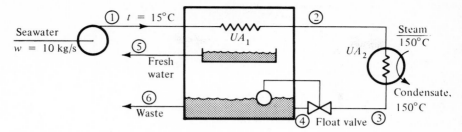

**Figure 6-16** Desalination plant in Prob. 6-3.

**6-3** A seawater desalination plant operates on the cycle shown in Fig. 6-16. Seawater is pressurized, flows through a heat exchanger, where its temperature is elevated by the condensation of what becomes the desalted water, and flows next through a steam heat exchanger, where it is heated but is still in a liquid state at point 3. In passing through the float valve the pressure drops and some of the liquid flashes into vapor, which is the vapor that condenses as fresh water. The portion at point 4 that remains liquid flows out as waste at point 6. The following conditions and relationships are known:

Temperature and flow rate of entering seawater.
$UA_1$ and $UA_2$ of the heat exchangers.
Enthalpies of saturated liquid and saturated vapor of seawater and the fresh water as functions of temperature:

$$h_f = f_1(t) \qquad \text{and} \qquad h_g = f_2(t)$$

For heat exchangers with one fluid condensing, use Eq. (5-10).
The system operates so that essentially $t_4 = t_5 = t_6$.

Set up an information-flow diagram that would be used for a successive-substitution system simulation, indicating which equations apply to each block. For convenience in checking, use these variables: $t_2$, $t_3$, $h_3$, $h_4$, $h_{f,4}$, $h_{g,4}$, $t_4$, $w_5$, $x_4$, and $q$ where $x_4$ is the fraction of vapor at point 4 and $q$ is the rate of heat transfer at the fresh-water condenser.

**6-4** In a synthetic-ammonia plant (Fig. 6-17) a $1:3$ mixture on a molar basis of $N_2$ and $H_2$ along with an impurity argon passes through a reactor where some of the nitrogen and hydrogen combine to form ammonia. The ammonia product formed leaves the system at the condenser and the remaining $H_2$, $N_2$, and Ar recycle to the reactor.

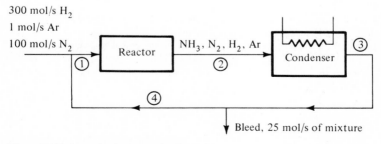

**Figure 6-17** A synthetic-ammonia plant.

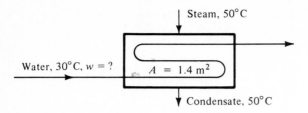

**Figure 6-18** Heat exchanger in Prob. 6-6.

The presence of the inert gas argon is detrimental to the reaction. If no argon is present, the reactor converts 60 percent of the incoming $N_2$ and $H_2$ into ammonia, but as the flow rate of argon through the reactor increases, the percent conversion decreases. The conversion efficiency follows the equation

$$\text{Conversion, } \% = 60e^{-0.016w}$$

where $w$ is the flow rate of argon through the reactor in moles per second.

To prevent the reaction from coming to a standstill, a continuous bleed of 25 mol/s of mixture of $N_2$, $H_2$, and Ar is provided. If the incoming feed consists of 100 mol/s of $N_2$, 300 mol/s of $H_2$, and 1 mol/s of Ar, simulate this system by successive substitution to determine the flow rate of mixture through the reactor and the rate of liquid ammonia production in moles per second.

**Ans.:** 893.2 and 188 mol/s.

**6-5** For $x(\tan x) = 2.0$, where $x$ is in radians, use the Newton-Raphson method to determine the value of $x$.

**Ans.:** 1.0769.

**6-6** The heat exchanger in Fig. 6-18 heats water entering at 30°C with steam entering as saturated vapor at 50°C and leaving as condensate at 50°C. The flow of water is to be chosen so that the heat exchanger transfers 50 kW. The area of the heat exchanger is 1.4 m², and the $U$ value of the heat exchanger based on this area is given by

$$\frac{1}{U} \text{ (m}^2 \cdot \text{K)/kW} = \frac{0.0445}{w^{0.8}} + 0.185$$

where $w$ is the flow rate of water in kilograms per second. Use the Newton-Raphson method for one equation and one unknown to determine the value of $w$ that results in the transfer of 50 kW.

**Ans.:** 0.6934 kg/s.

**6-7** Solve Prob. 6-1 using the multiple-equation Newton-Raphson method with trial values of SP = 200 Pa and $Q = 10$ m³/s.

**6-8** Air at 28°C with a flow rate of 4 kg/s flows through a cooling coil counterflow to cold water that enters at 6°C, as shown in Fig. 6-19. Air has a specific heat, $c_p = 1.0$ kJ/(kg · K). No dehumidification of the air occurs as it passes through the coil. The product of the area and heat-transfer coefficient for the heat exchanger is 7 kW/K. A pump just overcomes the pressure drop through the control valve and coil, such that $p_1 = p_4$. The pressure-flow characteristics of the pump are

$$p_2 - p_1, \text{Pa} = 120{,}000 - 15{,}400w_w^2$$

where $w_w$ is the flow rate of water in kilograms per second. The specific heat of the water is $c_p = 4.19$ kJ/(kg · K). The pressure drop through the coil is $p_3 - p_4 = 9260w_w^2$. The outlet-air temperature regulates the control valve to maintain an outlet-air temperature somewhere

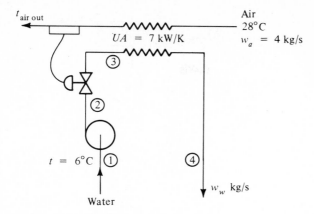

**Figure 6-19** Cooling coil in Prob. 6-8.

between 10 and 12°C. The flow–pressure-drop relation for the valve is $w_w = c_v \sqrt{p_2 - p_3}$, where $c_v$ is a function of the degree of valve opening, a linear relation, as shown in Fig. 6-20. The fully open value of $c_v$ is 0.012.

Using the Newton-Raphson method, simulate this system determining at least the following variables: $w_w$, $t_4$, $t_{air\ out}$, $p_2$, $p_3$, and $c_v$. Use as the test for convergence that the absolute values of pressures change less than 1.0, absolute values of temperatures less than 0.001, and absolute value of $c_v$ less than 0.000001. Limit the number of iterations to 10.

**Ans.:** $p_2 = 64{,}355$ (based on $p_1 = 0$), $t_4 = 14.14$, $c_v = 0.0108$.

**6-9** A two-stage air compressor with intercooler shown in Fig. 6-21 compresses air (which is assumed dry) from 100 to 1200 kPa absolute. The following data apply to the components:

$$\text{Displacement rate} = \begin{cases} 0.2 \text{ m}^3/\text{s} & \text{low-stage compressor} \\ 0.05 \text{ m}^3/\text{s} & \text{high stage compressor} \end{cases}$$

$$\text{Volumetric efficiency } \eta, \% = \frac{\text{flow rate measured at compressor suction, m}^3/\text{s}}{\text{displacement rate, m}^3/\text{s}} (100)$$

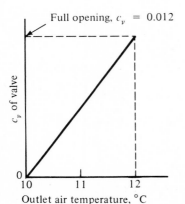

Full opening, $c_v = 0.012$

$c_v$ of valve

Outlet air temperature, °C

**Figure 6-20** Characteristics of valve in Prob. 6-8.

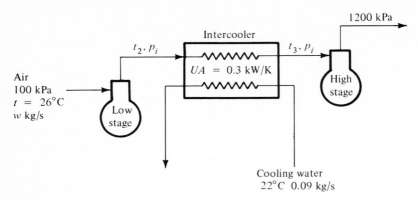

**Figure 6-21** Two-stage air compression in Prob. 6-9.

and for both compressors

$$\eta, \% = 104 - 4.0 \left( \frac{p_{disch}}{p_{suction}} \right)^{1.4}$$

The polytropic exponent $n$ in the equation $p_1 v_1^n = p_2 v_2^n$ is 1.2. The intercooler is a counter-flow heat exchanger receiving 0.09 kg/s of water at 22°C. The product of the overall heat-transfer coefficient and the area of this heat exchanger is $UA = 0.3$ kW/K. Assume that the air is a perfect gas.

Use the Newton-Raphson method to simulate this system, determining at least the values of $w$, $p_i$, $t_2$, and $t_3$. Use as a test for convergence that all variables change less than 0.001 during an iteration. Limit the number of iterations to 10.

**Ans.:** $w = 0.18$ kg/s, $t_2 = 101.8$°C, $p_i = 387.7$ kPa, $t_3 = 43.84$°C.

**6-10** A helium liquefier operating according to the flow diagram shown in Fig. 6-22 receives high-pressure helium vapor, liquefies a fraction of the vapor, and returns the remainder to be recycled. The following operating conditions prevail:

Point 1 (vapor entering warm side of heat exchanger), $T = 15$ K, $h = 78.3$ kJ/kg, $w = 5$ g/s, $p = 2000$ kPa

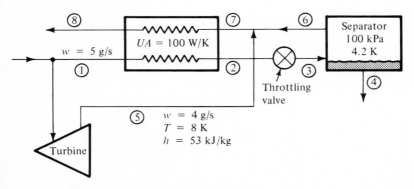

**Figure 6-22** Helium liquefier in Prob. 6-10.

Point 5 (vapor leaving turbine), $T = 8$ K, $h = 53$ kJ/kg, $w = 4$ g/s

Separator, $p = 100$ kPa, saturation temperature at 100 kPa = 4.2 K, $h_f = 10$ kJ/kg, $h_g = 31$ kJ/kg

Heat exchanger, $UA = 100$ W/K

Specific heat of helium vapor:

$$c_p = \begin{cases} 6.4 \text{ kJ/(kg} \cdot \text{K)} & \text{at 2000 kPa} \\ 5.8 \text{ kJ/(kg} \cdot \text{K)} & \text{at 100 kPa} \end{cases}$$

Using the Newton-Raphson method, simulate this system, determining the values of $w_4$, $T_2$, $T_7$, and $T_8$. Use as the test for convergence that all variables change less than 0.001 during an iteration. Limit the number of iterations to 10.

Ans.: $w_4 = 0.447$ g/s, $T_2 = 7.316$, $T_7 = 5.97$, and $T_8 = 10.93$ K.

**6-11** A refrigeration plant that operates on the cycle shown in Fig. 6-23 serves as a water chiller. Data on the individual components are as follows:

$$UA = \begin{cases} 30,600 \text{ W/K} & \text{evaporator} \\ 26,500 \text{ W/K} & \text{condenser} \end{cases} \qquad \text{Rate of water flow} = \begin{cases} 6.8 \text{ kg/s} & \text{evaporator} \\ 7.6 \text{ kg/s} & \text{condenser} \end{cases}$$

The refrigeration capacity of the compressor as a function of the evaporating and condensing temperatures $t_e$ and $t_c°$C, respectively, is given by the equation developed in Prob. 4-9

$$q_e, \text{kW} = 239.5 + 10.073t_e - 0.109t_e^2 - 3.41t_c - 0.00250t_c^2 - 0.2030t_e t_c$$
$$+ 0.00820t_e^2 t_c + 0.0013t_e t_c^2 - 0.000080005t_e^2 t_c^2$$

The compression power is expressed by the equation

$$P, \text{kW} = -2.634 - 0.3081t_e - 0.00301t_e^2 + 1.066t_c - 0.00528t_c^2 - 0.0011t_e t_c$$
$$- 0.000306t_e^2 t_c + 0.000567t_e t_c^2 + 0.0000031t_e^2 t_c^2$$

The condenser must reject the energy added in both the evaporator and the compressor. Determine the values of $t_e$, $t_c$, $q_e$, and $P$ for the following combinations of inlet water

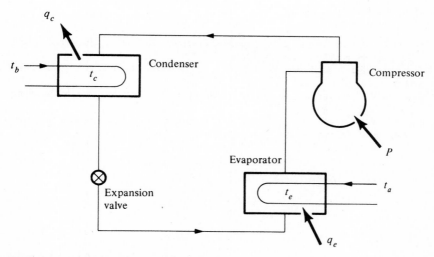

**Figure 6-23** Refrigeration plant in Prob. 6-11.

temperatures:

| Evaporator inlet water $t_a$, °C | 10 | 15 | 10 | 15 |
|---|---|---|---|---|
| Condenser inlet water $t_b$, °C | 25 | 25 | 30 | 30 |

Continue iterations until all variables change by an absolute value less than 0.1 percent during an iteration.

**Ans.:** For $t_a = 10°C$ and $t_b = 25°C$, $t_e = 2.84°C$, $t_c = 34.05°C$, $q_e = 134.39$ kW and $P = 28.34$ kW.

# REFERENCES

1. G. M. Dusinberre and J. C. Lester, *Gas Turbine Power*, International Textbook, Scranton, Pa., 1958.
2. *Procedures for Simulating the Performance of Components and Systems for Energy Calculations*, American Society of Heating, Refrigerating, and Air-Conditioning Engineers, New York, 1975.
3. H. A. Mosler, "PACER—A Digital Computer Executive Routine for Process Simulation and Design," M.S. thesis, Purdue University, Lafayette, Ind., January 1964.
4. C. M. Crowe, A. E. Hamielec, T. W. Hoffman, A. I. Johnson, D. R. Woods, and P. T. Shannon, *Chemical Plant Simulation; an Introduction to Computer-Aided Steady-State Process Analysis*, Prentice-Hall, Englewood Cliffs, N.J., 1971.
5. W. F. Stoecker, "A Generalized Program for Steady-State System Simulation," *ASHRAE Trans.*, vol. 77, pt. I, pp. 140–148, 1971.

# ADDITIONAL READINGS

Chen, C-C., and L. B. Evans: "More Computer Programs for Chemical Engineers," *Chem. Eng.*, vol. 86, no. 11, pp. 167–173, May 21, 1979.

Henley, E. J., and E. M. Rosen: *Material and Energy Balance Computations*, Wiley, New York, 1969.

Naphtali, L. M., "Process Heat and Material Balances," *Chem. Eng. Prog.*, vol. 60, no. 9, pp. 70–74, September 1964.

Peterson, J. N., C. Chen, and L. B. Evans: "Computer Programs for Chemical Engineers: 1978," pt. 1, *Chem. Eng.*, vol. 85, no. 13, pp. 145–154, June 5, 1978.

# SEVEN
## OPTIMIZATION

## 7-1 INTRODUCTION

Optimization is the process of finding the conditions that give maximum or minimum values of a function. Optimization has always been an expected role of engineers, although sometimes on small projects the cost of engineering time may not justify an optimization effort. Often a design is difficult to optimize because of its complexity. In such cases, it may be possible to optimize subsystems and then choose the optimum combination of them. There is no assurance, however, that this procedure will lead to the true optimum.

Chapter 1 pointed out that in designing a workable system the process often consists of arbitrarily assuming certain parameters and selecting individual components around these assumptions. In contrast, when optimization is an integral part of the design, the parameters are free to float until the combination of parameters is reached which optimizes the design.

Basic to any optimization process is the decision regarding which criterion is to be optimized. In an aircraft or space vehicle, minimum weight may be the criterion. In an automobile, the size of a system may be the criterion. Minimum cost is probably the most common criterion. On the other hand, the minimum owning and operating cost, even including such factors as those studied in Chap. 3 on economics, may not always be followed strictly. A manufacturer of domestic refrigerators, for example, does not try to design his system to provide minimum total cost to the consumer during the life of the equipment. The achievement of minimum first cost, which enhances sales, is more important than operating cost, although the operating cost cannot be completely out of bounds. Industrial organizations often turn aside from the most economical solution by introducing human, social, and aesthetic concerns. What is happening is that their criterion function includes not only monetary factors but also some other factors that may admittedly be only vaguely defined.

Optimization activities are often practiced under the name of *operations research*. Many developments in operations research emerged from attempts to optimize mathematical models of economic systems. It is only recently that mechanical and chemical engineers have used certain of the disciplines to optimize fluid- and energy-flow systems.

Component simulation and system simulation are often preliminary steps to optimizing thermal systems, since it may be necessary to simulate the performance over a wide range of operating conditions. A system that may be optimum for design loads may not be optimum over the entire range of its expected operation.

## 7-2 LEVELS OF OPTIMIZATION

Sometimes a design engineer will say: I have optimized the design by examining four alternate concepts to do the job, which probably means that the engineer has compared *workable* systems of four different concepts. The statement does emphasize the two levels of optimization, comparison of alternate concepts and optimization within a concept. All the optimization methods presented in the following chapters are optimizations within a concept. The flow diagram and mathematical representation of the system must be available at the beginning, and the optimization process consists of a give and take of sizes of individual components. All this optimization is done within a given concept. There is nothing in the upcoming procedures that will jump from one model to a better one. No optimization procedure will automatically shift the system under consideration from a steam-electric generating plant to a fuel-cell concept, for example.

A complete optimization procedure, then, consists of proposing all reasonable alternate concepts, optimizing the design of each concept, and then choosing the best of the optimized designs.

## 7-3 MATHEMATICAL REPRESENTATION OF OPTIMIZATION PROBLEMS

The elements of the mathematical statement of optimization include specification of the function and the constraints. Let $y$ represent the function that is to be optimized, called the *objective function; y* is a function of $x_1, x_2, \ldots, x_n$, which are called the *independent variables*. The objective function, then, is

$$y = y(x_1, x_2, \ldots, x_n) \longrightarrow \text{optimize} \qquad (7-1)$$

In many physical situations there are constraints, some of which may be equality constraints

$$\phi_1 = \phi_1(x_1, x_2, \ldots, x_n) = 0 \qquad (7\text{-}2)$$

$$\cdots\cdots\cdots\cdots\cdots\cdots\cdots\cdots\cdots$$

$$\phi_m = \phi_m(x_1, x_2, \ldots, x_n) = 0 \qquad (7\text{-}3)$$

as well as inequality constraints

$$\psi_1 = \psi_1(x_1, x_2, \ldots, x_n) \leqslant L_1 \qquad (7\text{-}4)$$

$$\cdots\cdots\cdots\cdots\cdots\cdots\cdots\cdots\cdots$$

$$\psi_j = \psi_j(x_1, x_2, \ldots, x_n) \leqslant L_j \qquad (7\text{-}5)$$

The physical conditions dictate the sense of the inequalities in Eqs. (7-4) to (7-5).

An additive constant appearing in the objective function does not affect the values of the independent variables at which the optimum occurs. Thus, if

$$y = a + Y(x_1, \ldots, x_n)$$

where $a$ is a constant, the minimum of $y$ can be written

$$\min\,[a + Y(x_1, \ldots, x_n)] = a + \min\,[Y(x_1, \ldots, x_n)] \qquad (7\text{-}6)$$

A further property of the optimum is that the maximum of a function occurs at the same state point at which the minimum of the negative of the function occurs, thus

$$\max\,[y(x_1, \ldots, x_n)] = -\min\,[-y(x_1, \ldots, x_n)] \qquad (7\text{-}7)$$

## 7-4 A WATER-CHILLING SYSTEM

A water-chilling system, shown schematically in Fig. 7-1, will be used to illustrate the mathematical statement. The requirement of the system is that it cool 20 kg/s of water from 13 to 8°C, rejecting the heat to the atmosphere through a cooling tower. We seek a system with a minimum first cost to perform this duty.

Designate the sizes of the components in the system by $x_{CP}$, $x_{EV}$, $x_{CD}$, $x_P$, and $x_{CT}$, which represent the sizes of the compressor, evaporator, condenser, pump, and cooling tower, respectively. The total cost $y$ is the sum of the individual first plus installation costs, and this is the quantity that we wish to minimize.

$$y(x_{CP}, x_{EV}, x_{CD}, x_P, x_{CT}) \longrightarrow \text{minimize} \qquad (7\text{-}8)$$

With only the statement of Eq. (7-8), the minimum could be achieved by shrinking the sizes of all components to zero. Overlooked is the requirement that the

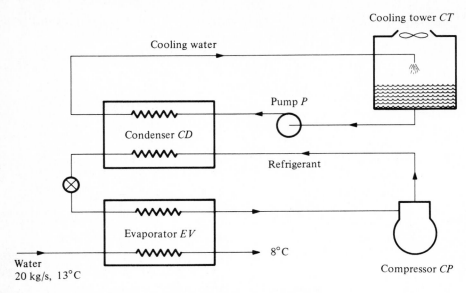

**Figure 7-1** Water-chilling unit being optimized for minimum first cost.

combination of sizes be such that the water-chilling assignment of providing

$$(20 \text{ kg/s}) (13 - 8°C) [4.19 \text{ kJ/(kg} \cdot \text{K)}] = 419 \text{ kW}$$

of refrigeration is accomplished. Equation (7-9) expresses this constraint

$$\phi(x_{CP}, x_{EV}, x_{CD}, x_P, x_{CT}) = 419 \text{ kW} \qquad (7\text{-}9)$$

where $\phi$ is understood to mean the cooling capacity as a function of component sizes when 20 kg/s of water enters at 13°C. Actually, Eq. (7-9) could be an inequality constraint, because probably no one would object to a larger capacity than the requirement of 419 kW if the cost were not increased.

Some practical considerations impose certain inequality constraints. The system should be designed so that the evaporating temperature $t_{ev}$ is above 0°C or, at the lowest, −2 to −1°C to prevent water from freezing on the tube surfaces. This constraint is

$$t_{ev}(x_{CP}, x_{EV}, x_{CD}, x_P, x_{CT}) \geqslant 0°C \qquad (7\text{-}10)$$

An extremely high discharge temperature $t_d$ of the refrigerant leaving the compressor may impair the lubrication

$$t_d(x_{CP}, x_{EV}, x_{CD}, x_P, x_{CT}) \leqslant 110°C \qquad (7\text{-}11)$$

There may be other inequality constraints, such as limiting the condenser cooling-water flow in relation to the size of the cooling tower to prevent it from splashing out.

The elements of the optimization problem are all present here, the objective function, equality constraints, and inequality constraints, all in terms of the independent variables, which are the sizes of components.

## 7-5 OPTIMIZATION PROCEDURES

In the next few sections several optimization methods will be listed. Although this list includes most of the frequently used methods in engineering practice, it is nowhere near exhaustive. In the optimization of systems, it is almost axiomatic that the objective function be dependent upon more than one variable. In fact, some thermal systems may have dozens or even hundreds of variables which demand sophisticated optimization techniques. While considerable effort may be required in the optimization process, developing mathematical relationships for the function to be optimized and the constraints may also require considerable effort.

## 7-6 CALCULUS METHODS: LAGRANGE MULTIPLIERS

The basis of optimization by calculus, presented in Chap. 8, is to use derivatives to indicate the optimum. The method of Lagrange multipliers performs an optimization where equality constraints exist but the method cannot directly accommodate inequality constraints. A necessary requirement for using calculus methods is the ability to extract derivatives of the objective function and constraints.

## 7-7 SEARCH METHODS

These methods, covered in Chap. 9, involve examining a number of combinations of values of the independent variables and drawing conclusions from the magnitude of the objective function at these combinations. An obvious possibility is to calculate the value of the function at all possible combinations of, for example, 20 values distributed through the range of interest of one parameter, each in combination with 20 of the second parameter, and so on. Such a search method is not very imaginative and is also inefficient. The search methods of interest are those which are efficient, particularly when applied to multivariable optimization.

When applying search methods to continuous functions, since only discrete points are examined, the exact optimum can only be approached, not reached, by a finite number of trials. On the other hand, when optimizing systems where the components are available only in finite steps of sizes, search methods are often superior to calculus methods, which assume an infinite gradation of sizes.

## 7-8 DYNAMIC PROGRAMMING

The word "programming" here and in the next several sections means optimization and has no direct relationship with computer programming, for example. This method of optimization, discussed in Chap. 10, is unique in that the result is an optimum function rather than an optimum state point. The result of the optimization of all the other methods mentioned here is a set of values of the independent variables $x_1$ to $x_n$ that result in the optimal value of the objective function $y$. The problem attacked by dynamic programming is one where the desired result is a *path*, e.g., the best route of a gas pipeline. The result is therefore a function relating several variables. Dynamic programming is related to the calculus of variations, and it does in a series of discrete processes what the calculus of variations does continuously.

## 7-9 GEOMETRIC PROGRAMMING

Probably the youngest member of the programming family is geometric programming, discussed in Chap. 11. Geometric programming optimizes a function that consists of a sum of polynomials wherein the variables may appear to integer and noninteger exponents. When the usefulness of polynomial expressions in Chap. 4 is recalled, it is clear that the form of the function to which geometric programming is applicable is one that occurs frequently in thermal systems.

## 7-10 LINEAR PROGRAMMING

Chapter 12 presents an introduction to linear programming, which is a widely used and well-developed discipline applicable when all the equations (7-1) to (7-5) are linear. The magnitude of problems now being solved by linear programming is enormous, occasionally extending into optimizations which contain several thousand variables.

## 7-11 SETTING UP THE MATHEMATICAL STATEMENT OF THE OPTIMIZATION PROBLEM

One of the first steps in performing an optimization is to translate the physical situation into a mathematical statement. The desired form is that comparable to Eqs. (7-1) to (7-5) because the optimization techniques can pick up the problem when the objective function and the constraints are specified. In the optimization of thermal systems, establishing the objective function is often simple and sometimes even a trivial task. The challenge usually arises in writing the constraints. A strategy that is often successful is to (1) specify all the direct constraints, such as requirements of capacity, limitations of temperature and pressure, etc., (2) describe in equation form the component characteristics and properties of working substances, and (3) write mass and energy balances. Operations (1) to (3) usually provide a set of equations containing more variables than exist in the objective function. The set of constraint equations is then reduced in number by eliminating variables that do not exist in the objective function. It should be pointed out, however, that some optimization techniques permit additional variables in the constraints that do not exist in the objective function. Steps (2) and (3) in setting up the constraints are similar to the process used in structuring a system-simulation problem.

**Example 7-1** Between two stages of air compression, the air is to be cooled from 95 to 10°C. The facility to perform this cooling, shown in Fig. 7-2, first cools the air in a precooler and then in a refrigeration unit. Water passes through the condenser of the refrigeration unit, then into the precooler, and finally to a cooling tower, where heat is rejected to the atmosphere.

The flow rate of compressed air is 1.2 kg/s, and the specific heat is 1.0 kJ/(kg · K). The flow rate of water is 2.3 kg/s, and its specific heat is 4.19 kJ/(kg · K). The water leaves the cooling tower at a temperature 24°C. The system is to be designed for minimum first cost, where this first cost comprises the cost of the refrigeration unit, precooler, and cooling tower, designated $x_1$, $x_2$, and $x_3$, respectively, in dollars. The expressions for these

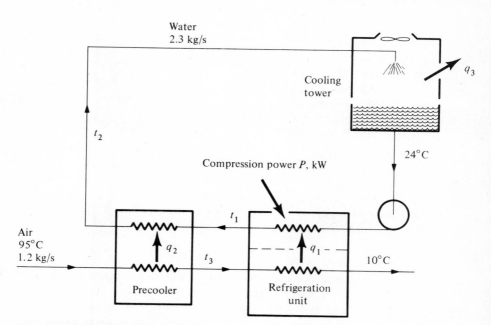

**Figure 7-2** Air-cooling system in Example 7-1.

costs are

Refrigeration unit:    $$x_1 = 48q_1 \qquad (7\text{-}12)$$

Precooler:    $$x_2 = \frac{50q_2}{t_3 - t_1} \qquad (7\text{-}13)$$

where the equation is applicable when $t_3 > t_1$

Cooling tower:    $$x_3 = 25q_3 \qquad (7\text{-}14)$$

where the $q$'s are rates of heat transfer in kilowatts, as designated in Fig. 7-2. The compression power $P$ kW required by the refrigeration unit is $0.25q_1$, and both $q_1$ and the compression power must be absorbed by the condenser cooling water passing through the refrigeration unit.

Develop ($a$) the objective function and ($b$) the constraint equations for an optimization to provide minimum first cost.

**Solution** The goal of this example is only to set up the optimization problem in the form of Eqs. (7-1) to (7-5) and not to perform the actual optimization. Before proceeding, however, it would be instructive to examine qualitatively the optimization features of this system. Since the precooler is a simple heat exchanger, under most operating conditions it is less costly for a given heat-transfer rate than the refrigeration unit. It would appear preferable, then, to do as much cooling of the air as possible with the precooler. However, as the temperature $t_3$ approaches the value of $t_1$, the size of the precooler becomes very large. Some capacity is required of the refrigeration unit in order to cool the air below 24°C. The cooling tower must reject all the heat from the system, which includes the heat from the air as well as the compression power to drive the refrigeration unit. Shifting more cooling load to the refrigeration unit increases the size and cost of the cooling tower moderately.

($a$) The first assignment is to develop the expression for the objective function. Since the total first cost is to be minimized, the objective function will be the first cost in terms of the variables of optimization. A choice must be made of these variables; the objective function could conceivably be written in terms of the costs of the individual components (the $x$'s), the energy flow rates (the $q$'s), or even the temperatures ($t_1$, $t_2$, and $t_3$). The most straightforward choice is to use the component costs

$$\text{Total cost} = y = x_1 + x_2 + x_3 \qquad (7\text{-}15)$$

If the $q$'s are chosen as the variables of optimization, it is necessary to start with Eq. (7-15) and express the $x$'s in terms of the $q$'s.

($b$) The next task is to write the constraints, and this means developing the set of equations in terms of the variables used in the objective function. Establishing the objective function is usually a simple process; the major challenge is setting up the constraints. The advice in the early part of this section was to specify the direct constraints, the component characteristics, and finally the energy and mass balances. This expanded set of equations is then reduced by eliminating the variables that do not appear in the objective function.

A direct constraint is the requirement that the airflow rate of 1.2 kg/s be cooled from 95 to 10°C. This requirement can be expressed in two equations

$$q_1 = (1.2 \text{ kg/s}) [1.0 \text{ kJ/(kg} \cdot \text{K)}] (t_3 - 10) \qquad (7\text{-}16)$$

$$q_2 = (1.2 \text{ kg/s}) [1.0 \text{ kJ/(kg} \cdot \text{K)}] (95 - t_3) \qquad (7\text{-}17)$$

Under the heading of component characteristics fall the expression for the compression power equaling $0.25q_1$ and the relationships of the sizes (costs) to the capacity, Eqs. (7-12) to (7-14).

The final category includes energy and mass balances

Refrigeration unit: $\qquad q_1 + P = (2.3 \text{ kg/s}) [4.19 \text{ kJ/(kg} \cdot \text{K)}] (t_1 - 24)$

Precooler: $\qquad (1.2) (1.0) (95 - t_3) = (2.3) (4.19) (t_2 - t_1)$

Cooling tower: $\qquad (2.3) (4.19) (t_2 - 24) = q_3$

The complete set of constraint equations is

$$q_1 = (1.2) (1.0) (t_3 - 10) \qquad (7\text{-}18)$$

$$q_2 = (1.2) (1.0) (95 - t_3) \qquad (7\text{-}19)$$

$$P = 0.25q_1 \qquad (7\text{-}20)$$

$$x_1 = 48q_1 \qquad (7\text{-}21)$$

$$x_2 = \frac{50q_2}{t_3 - t_1} \qquad (7\text{-}22)$$

$$x_3 = 25q_3 \qquad (7\text{-}23)$$

$$q_1 + P = (2.3) (4.19) (t_1 - 24) \qquad (7\text{-}24)$$

$$(1.2) (1.0) (95 - t_3) = (2.3) (4.19) (t_2 - t_1) \qquad (7\text{-}25)$$

$$(2.3) (4.19) (t_2 - 24) = q_3 \qquad (7\text{-}26)$$

There are nine equations in the set, Eqs. (7-18) to (7-26) and ten unknowns, $q_1, q_2, q_3, P, x_1, x_2, x_3, t_1, t_2$, and $t_3$. The next operation is to eliminate in this set of equations all but the variables of optimization, $x_1, x_2$, and $x_3$. As the elimination of variables and equations proceeds, there will always be one more unknown than the number of equations, so when all but the three $x$'s are eliminated, there should be two equations remaining. These two constraint equations are Eqs. (7-28) and (7-29), so the complete mathematical statement of this optimization problem is:

Minimize
$$y = x_1 + x_2 + x_3 \qquad (7\text{-}27)$$

subject to

$$0.01466 x_1 x_2 - 14 x_2 + 1.042 x_1 = 5100 \qquad (7\text{-}28)$$

$$7.69 x_3 - x_1 = 19{,}615 \qquad (7\text{-}29)$$

## 7-12 DISCUSSION OF EXAMPLE 7-1

The objective of this chapter is to introduce procedures for setting up the mathematical statement of the optimization problem. With one of the methods in the subsequent chapters the execution of the optimization process would show that the optimal values of $x_1$, $x_2$, and $x_3$ are \$1450, \$496, and \$2738, respectively. Equation (7-13) [or Eq. (7-22)] was presented as valid if $t_3 > t_1$, and this condition could legitimately be listed as one of the constraints. If $t_3 < t_1$, $x_2$ becomes negative, which is physically impossible.

The constraints are an integral part of the statement of the optimization problem. The objective function without the constraints is meaningless because the $x$'s could all shrink to zero and there would be no cost for the system. The constraint in Eq. (7-28) requires a positive value of $x_1$ which is the same as requiring the existence of a refrigeration unit. From heat-transfer considerations the precooler can cool the air no lower than a temperature of 24°C. Substituting $x_1 = 0$ in Eq. (7-28) makes $x_2$ negative, which is physically impossible. Equation (7-28) does permit $x_2$ to be zero, in which case all cooling is performed by the refrigeration unit.

The constraint equation (7-29) imposes a minimum value of the cooling-tower size and cost $x_3$. As the size of the refrigeration unit and $x_1$ increases, $x_3$ also increases because of the compression power associated with the refrigeration unit.

## 7-13 SUMMARY

While it is true that engineers have always sought to optimize their designs, it has been only since the widespread application of the digital computer that sophisticated methods of optimization have become practical for complex systems. The application of optimization techniques to large-scale thermal systems is still in its infancy, but experience so far indicates that setting up the problem to the point where an optimization method can take over represents perhaps 70 percent of the total effort. The emphasis on optimization techniques in the next five chapters may suggest that engineers are home free once they know several methods. Realistically, however, the execution of the optimization can only begin when the characteristics of the physical system have been converted into the equations for the objective function and constraints.

## PROBLEMS

**7-1** The flow rate of raw material to the processing plant shown in Fig. 7-3 is 10 kg/s of mixture consisting of 50 percent A and 50 percent B. The separator can remove some of material A in pure form, and this product has a selling price of $10 per kilogram. The other product from the separator sells for $2 per kilogram, and some of this stream may be recycled. The cost of operating the separator in dollars per second is

$$\frac{0.2w_1}{1 - (w_2/w_1 x_1)}$$

where $w$ = mass rate of flow, kg/s
$x$ = fraction of A in the mixture

(This separator cost equation indicates that the cost becomes infinite for perfect separation.)

(a) Set up the objective function in terms of $w_1$, $w_2$, and $x_1$ to maximize the profit. The cost of the raw material is constant.

(b) Set up the constraint(s) in terms of $w_1$, $w_2$, and $x_1$.

Ans.: (b) $(5 - w_1 x_1)(w_1 - w_2) = (10 - w_1)(w_1 x_1 - w_2)$.

**7-2** The plant shown schematically in Fig. 7-4 produces chlorine by the reaction of HCl and $O_2$ according to the equation

$$4HCl + O_2 \longrightarrow 2H_2O + 2Cl_2$$

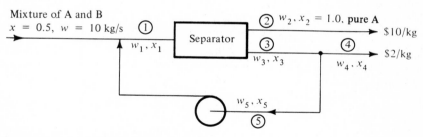

**Figure 7-3** Processing plant in Prob. 7-1.

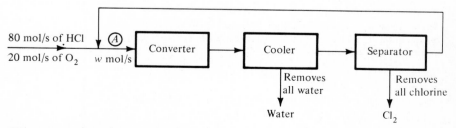

**Figure 7-4** Chlorine plant in Prob. 7-2.

The plant receives 80 mol/s of HCl and 20 mol/s of $O_2$. The converter is capable of achieving only a partial conversion, and the unreacted HCl and $O_2$ are recycled following removal of the water and the chlorine.

The first cost of the system is to be minimized, and the major variable affecting the cost is that of the converter whose cost is represented by

$$\text{Cost per mol/s of reactants at } A, \text{ dollars} = 800 + \frac{24,000}{100 - x}$$

where $x$ = conversion efficiency, %
 or $x$ = (fraction of entering HCl and $O_2$ that reacts) $\times$ 100

Determine (a) the objective function in terms of $w$ and $x$ and (b) the constraint(s) that permit optimization for minimum cost.

Ans.: (b) $wx = 10,000$.

**7-3** A supersonic wind-tunnel facility is being designed in which the air will flow in series through a compressor, storage tank, pressure-control valve, the wind tunnel, and thence to exhaust. During tests a flow rate of 5 kg/s must be available to the wind tunnel at a pressure of 400 kPa. The tests are intended to study heat transfer, and before each test 120 s of stabilization time is required, during which 5 kg/s must also flow. A total of 3600 s of useful test time is required during an 8-h period, and this 3600 s can be subdivided into any number of equal-length tests.

The mode of operation is to start the compressor and allow it to run continuously at full capacity during the 8-h period. Each cycle consists of the following stages: (1) buildup of pressure in the storage tank from 400 to 530 kPa, during which time there is no flow through the wind tunnel, (2) stabilization period of 120 s, during which the flow is 5 kg/s and the storage tank pressure begins to drop, and (3) the useful test, during which flow is 5 kg/s and at the end of which the storage-tank pressure has dropped to 400 kPa.

A pressure of 400 kPa is available in the storage tank at the start of the day.

The compressor–storage-tank combination is to be selected for minimum total first cost. The compressor cost in dollars is given by

$$\text{Cost} = 800 + 2400S$$

where $S$ is the capacity of the compressor in kilograms per second. The storage tank is to be a cube, for which metal costs $15 per square meter. The mass of air that can be stored in the tank at 530 kPa in excess of that at 400 kPa is $1.5V$ kg, where $V$ is the volume in cubic meters.

(a) Write the expression for the total first cost of the compressor and tank in terms of $S$ and $V$.

(b) Develop the constraint equation in terms of $S$ and $V$ to meet the operating conditions.

Ans.: (b) $S\left(1 + \dfrac{1.5V}{137S + 0.214V}\right) = 5.$

**7.4** A simplified version of a combined gas- and steam-turbine plant[1] for a liquefied-petroleum-gas facility at Bushton, Kansas is shown in Fig. 7-5; the plant must meet the following minimum requirements:

|  | kW |
| --- | --- |
| Power to propane compressor | 3800 |
| Low-pressure steam equivalent for process use | 6500 |
| High-pressure steam equivalent for process use | 8800 |

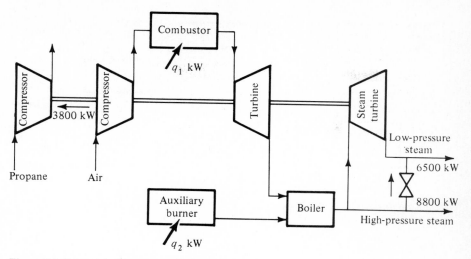

**Figure 7-5** Schematic diagram of system in Prob. 7-4.

In the gas-turbine plant, 20 percent of the heating value of natural gas is converted into mechanical power and the remaining 80 percent passes to the exhaust gas. As the exhaust gas flows through the boiler, 60 percent of its heat is converted into steam. The boiler is also equipped with an auxiliary burner, which permits 80 percent of the heating value of the natural gas to be converted into steam. High-pressure steam flows to process use and/or to the steam turbine, where 15 percent of the thermal energy is converted into mechanical power.

The heat input rates from the natural gas at the gas turbine and the boiler are designated $q_1$ and $q_2$ kW, respectively.

The plant is to be operated in such a way that it consumes a minimum total quantity of natural gas. In terms of $q_1$ and $q_2$ (a) write the objective function and (b) develop the constraint equations. Acknowledge by inequalities in the constraint equations the possibility of dumping power or steam.

Ans.: (b) $q_1 + 1.176q_2 \geqslant 28{,}090$ kW

$q_1 + 0.441q_2 \geqslant 18{,}820$ kW

$q_1 + 1.665q_2 \geqslant 31{,}880$ kW

7-5 Nagib[2] discusses improvement in the operating efficiency of a gas-turbine cycle by precooling the air entering the compressor. The precooler is an absorption-refrigeration unit, which is supplied with steam generated from the heat of the exhaust gas, as shown in Fig. 7-6. Achievement of this improved efficiency entails additional investment cost for the absorption unit and boiler. Furthermore, there is an optimum combination of absorption unit, boiler, and regenerator.

### Performance data

Compressor power, kW = $34(t_1 + 273)$.

Absorption unit delivers 0.6 kJ of cooling per kilojoule of heat supplied by the steam.

The flow rate of gas through the cycle is 45 kg/s, and negligible addition of mass at the combustor is assumed.

$U$ value of the boiler based on steam-side area, 0.25 kW/(m² · K).

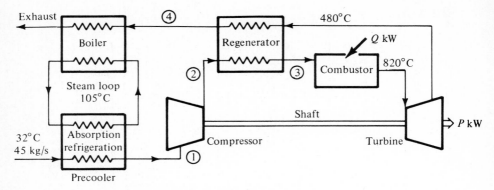

**Figure 7-6** Precooler of gas-turbine inlet air in Prob. 7-5.

$U$ value of regenerator, 0.082 kW/(m² · K).
$c_p$ of gas throughout the cycle, 1.0 kJ/(kg · K).
Assume no heat transfer to ambient from any component.
Additional temperatures are indicated on Fig. 7-6.

*Cost data*

Present worth of power generated during the life of the plant, $225P$ dollars
Present worth of fuel cost for the plant life, $45Q$ dollars
First cost of absorption unit, $80A$ dollars
First cost of boiler, $1,000B$ dollars
First cost of regenerator, $90C$ dollars

where $Q$ = heat rate at combustor, kW
$P$ = power generated, kW
$A$ = absorption unit size, kW of cooling
$B$ = boiler size, m² of steam-side area
$C$ = regenerator size, m²

(a) Develop the objective function for the total present worth of the profit of the system in terms of $A, B, C, Q,$ and $P$.

(b) Develop expressions for the temperatures $t_1, t_2, t_3,$ and $t_4$ in terms of the variables of part (a).

(c) Develop the constraint equations in terms of the variables of part (a).

**Ans.:** (b): $t_1 = 32 - 0.0222A$ $t_2 = 262.5 - 0.039A$

$$t_3 = 262.5 - 0.039A + \frac{217.5 + 0.039A}{548.8/C + 1}$$

$$t_4 = 480 - \frac{217.5 + 0.039A}{548.8/C + 1}$$

(c): $P - 4930 - 0.7548A = 0$

$$A - (27)\left(375 - \frac{217.5 + 0.039A}{548.8/C + 1}\right)(1 - e^{-0.005556B}) = 0$$

$$Q - (45)\left(557.5 + 0.039A - \frac{217.5 + 0.039A}{548.8/C + 1}\right) = 0$$

# REFERENCES

1. B. G. Wobker and C. E. Knight, "Mechanical Drive Combined-Cycle Gas and Steam Turbines for Northern Gas Products," *ASME Pap.* 67-GT-39, 1967.
2. M. M. Nagib, "Analysis of a Combined Gas Turbine and Absorption–Refrigeration Cycle," *ASME Pap.* 70-PWR-18, 1970.

# ADDITIONAL READINGS

Beveridge, G. S. G., and R. S. Schechter: *Optimization: Theory and Practice*, McGraw-Hill, New York, 1970.

Denn, M. M.: *Optimization by Variational Methods*, McGraw-Hill, New York, 1969.

Fox, R. L.: *Optimization Methods for Engineering Design*, Addison-Wesley, Reading, Mass., 1971.

Ray, W. H., and J. Szekely: *Process Optimization*, Wiley, New York, 1973.

Rosenbrock, H. H., and C. Storey: *Computational Techniques for Chemical Engineers*, Pergamon, New York, 1966.

Wilde, D. J., and C. S. Beightler: *Foundations of Optimization*, Prentice-Hall, Englewood Cliffs, N.J., 1967.

# EIGHT

## LAGRANGE MULTIPLIERS

### 8-1 CALCULUS METHODS OF OPTIMIZATION

Classical methods of optimization are based on calculus and specifically determine the optimum value of a function as indicated by the nature of the derivatives. In order to optimize using calculus, the function must be differentiable and any constraints must be equality constraints. That there is a need for any method other than calculus, such as linear and nonlinear programming, may arise from the appearance of inequality constraints. In addition, the fact that the function is not continuous but exists only at specific values of the parameters rules out calculus procedures and favors such techniques as search methods. On the other hand, since some of the operations in the calculus methods appear in slightly revised forms in the other optimization methods, calculus methods are important both for the cases that they can solve in their own right and to illuminate some of the procedures in noncalculus methods.

This chapter presents the Lagrange multiplier equations, proceeds to the interpretation and mechanics of optimization using these equations, and then begins the presentation of the background and visualization of constrained optimization. Since unconstrained optimization is only a special case of constrained optimization, the method of Lagrange multipliers is applicable to all situations explored in this chapter. This chapter also explains how an optimum condition can be tested to establish whether the condition is a maximum or a minimum and, finally, introduces the concept of the sensitivity coefficient.

## 8-2 THE LAGRANGE MULTIPLIER EQUATIONS

The mathematical statement of the constrained optimization problem first given in Eqs. (7-1) to (7-3) is repeated here:

Optimize
$$y = y(x_1, x_2, \ldots, x_n) \tag{8-1}$$

subject to
$$\phi_1(x_1, x_2, \ldots, x_n) = 0 \tag{8-2}$$
$$\cdots\cdots\cdots\cdots\cdots\cdots$$
$$\phi_m(x_1, x_2, \ldots, x_n) = 0 \tag{8-3}$$

The method of Lagrange multiplers states that the optimum occurs at values of the $x$'s that satisfy the equations

$$\nabla y - \lambda_1 \nabla\phi_1 - \cdots - \lambda_m \nabla\phi_m = 0 \tag{8-4}$$
$$\phi_1(x_1, \ldots, x_n) \quad = 0 \tag{8-5}$$
$$\cdots\cdots\cdots\cdots\cdots\cdots\cdots$$
$$\phi_m(x_1, \ldots, x_n) \quad = 0 \tag{8-6}$$

The remainder of the chapter explains the meaning of the symbols and operations designated in the equations, the mechanics of solving the equations, and applications, examples, and geometric visualization of the Lagrange multipler equations.

## 8-3 THE GRADIENT VECTOR

The inverted delta designates the *gradient vector*. A *scalar* is a quantity with a magnitude but no direction, while a *vector* has both magnitude and direction. By definition, the gradient of a scalar is

$$\nabla y = \frac{\partial y}{\partial x_1} i_1 + \frac{\partial y}{\partial x_2} i_2 + \cdots + \frac{\partial y}{\partial x_n} i_n \tag{8-7}$$

where $i_1, i_2, \ldots, i_n$ are unit vectors, which means that they have direction and their magnitudes are unity.

Suppose, for example, that a solid rectangular block has a temperature distribution that can be expressed in terms of the coordinates $x_1, x_2$, and $x_3$, as shown in Fig. 8-1. If the temperature $t$ is the following function of $x_1, x_2$, and $x_3$,

$$t = 2x_1 + x_1 x_2 + x_2 x_3^2$$

then the gradient of $t$, $\nabla t$, following the definition of Eq. (8-7), is

$$\nabla t = (2 + x_2) i_1 + (x_1 + x_3^2) i_2 + (2x_2 x_3) i_3$$

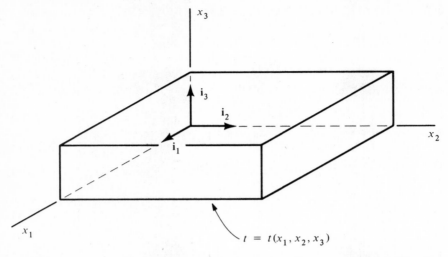

**Figure 8-1** Solid object in which a scalar, the temperature, is expressed as a function of $x_1$, $x_2$, and $x_3$.

where $i_1$, $i_2$, and $i_3$ are the unit vectors in the $x_1$, $x_2$, and $x_3$ directions, respectively. The gradient operation is one that converts a scalar quantity into a vector quantity.

## 8-4 FURTHER EXPLANATION OF LAGRANGE MULTIPLIER EQUATIONS

Also appearing in Eq. (8-4) is a series of lambdas, $\lambda_1, \ldots, \lambda_m$. These terms are constants and are the *Lagrange multipliers*, whose values are not known until the set of equations is solved.

Equation (8-4) is a vector equation, which means that for the terms on the left side of the equation to equal zero the coefficients of each unit vector must sum to zero. The vector equation (8-4) is thus a shorthand form for $n$ scalar equations,

$$i_1: \qquad \frac{\partial y}{\partial x_1} - \lambda_1 \frac{\partial \phi_1}{\partial x_1} - \cdots - \lambda_m \frac{\partial \phi_m}{\partial x_1} = 0 \qquad (8\text{-}8)$$

. . . . . . . . . . . . . . . . . . . . . . . . . . .

$$i_n: \qquad \frac{\partial y}{\partial x_n} - \lambda_1 \frac{\partial \phi_1}{\partial x_n} - \cdots - \lambda_m \frac{\partial \phi_m}{\partial x_n} = 0 \qquad (8\text{-}9)$$

These $n$ scalar equations, along with the $m$ constraint equations (8-5) to (8-6), form the set of $n + m$ simultaneous equations needed to solve for the same number of unknowns: $x_1^*, x_2^*, \ldots, x_n^*$ and $\lambda_1, \lambda_2, \ldots, \lambda_m$. The asterisk on the $x$'s is often used to indicate the values of the variables at the optimum. The optimal values of the $x$'s can be substituted into the objective function, Eq. (8-1), to determine the optimal value of $y$, designated $y^*$.

The number of equality constraints $m$ is always less than the number of variables $n$. In the limiting case where $m = n$, the constraints (if they are independent equations) fix the values of the $x$'s, and no optimization is possible.

## 8-5 UNCONSTRAINED OPTIMIZATION

The Lagrange multiplier equations can be used to attack constrained optimization problems, but the equations apply equally well to unconstrained optimization. The unconstrained optimization is a special (and simpler) case of constrained optimization. The objective function $y$ is a function of variables $x_1, \ldots, x_n$

$$y = y(x_1, \ldots, x_n) \tag{8-10}$$

When the Lagrange multiplier equations are applied to Eq. (8-10), since there are no $\phi$'s, the condition for optimum is

$$\nabla y = 0 \tag{8-11}$$

or
$$\frac{\partial y}{\partial x_1} = 0, \frac{\partial y}{\partial x_2} = 0, \ldots, \frac{\partial y}{\partial x_n} = 0 \tag{8-12}$$

The state point where the derivatives are zero is called a *critical point*, and it may be a maximum or minimum (one of which we are seeking), or it may be a saddle point or a ridge or valley. Further mathematical analysis may be necessary to determine the type of critical point, although in physical situations the nature of the point will often be obvious. We shall assume in the remainder of this chapter that Eqs. (8-12) describe a maximum or minimum.

A function $y$ of two variables $x_1$ and $x_2$ can be represented graphically as in Fig. 8-2. The minimum exists where

$$\frac{\partial y}{\partial x_1} = 0 = \frac{\partial y}{\partial x_2}$$

**Example 8-1** Determine the minimum value of $y$ where

$$y = x_1 x_2 + \frac{1}{x_1 x_3} - 16x_2^2 + x_3$$

SOLUTION The Lagrange multiplier equations for this unconstrained optimization are

$$\nabla y = 0 \quad \text{or} \quad \frac{\partial y}{\partial x_1} = \frac{\partial y}{\partial x_2} = \frac{\partial y}{\partial x_3} = 0$$

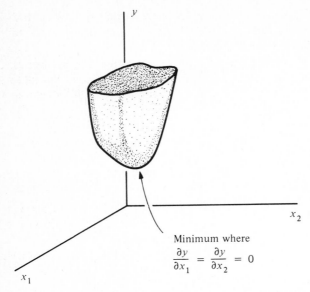

**Figure 8-2** Unconstrained optimum occurs where partial derivatives equal zero.

$$\frac{\partial y}{\partial x_1} = x_2 - \frac{1}{x_1^2 x_3} = 0$$

$$\frac{\partial y}{\partial x_2} = x_1 - 32x_2 = 0$$

$$\frac{\partial y}{\partial x_3} = -\frac{1}{x_1 x_3^2} + 1 = 0$$

Solving simultaneously gives $x_1^* = 4$, $x_2^* = \frac{1}{8}$, and $x_3^* = \frac{1}{2}$. Substitution of these values into the objective function yields

$$y^* = \frac{5}{4}$$

Sometimes it is convenient to convert a constrained problem into an unconstrained optimization by solving for a variable in a constraint equation and substituting that expression wherever else in the remaining constraints or objective function the variable appears. The number of constraints is progressively reduced until only the unconstrained objective function remains.

**Example 8-2** A total length of 100 m of tubes must be installed in a shell-and-tube heat exchanger (Fig. 8-3), in order to provide the necessary heat-transfer area. The total cost of the installation in dollars includes

1. The cost of the tubes, which is constant at \$900

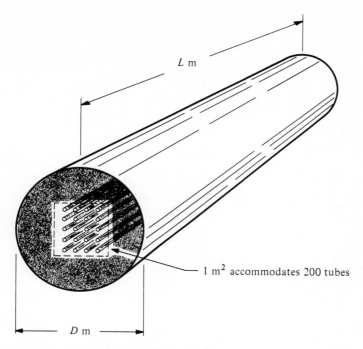

$L$ m

1 m² accommodates 200 tubes

$D$ m

**Figure 8-3** Heat exchanger in Example 8-2.

2. The cost of the shell = $1100D^{2.5}L$
3. The cost of the floor space occupied by the heat exchanger = $320DL$, where $L$ is the length of the heat exchanger and $D$ is the diameter of the shell, both in meters

The spacing of the tubes is such that 200 tubes will fit in a cross-sectional area of 1 m² in the shell.

Determine the diameter and length of the heat exchanger for minimum first cost.

SOLUTION  The objective function includes the three costs,

$$\text{Cost} = 900 + 1100D^{2.5}L + 320DL$$

The constraint requires the heat exchanger to include 100 m of tubes

$$\left(\frac{\pi D^2}{4} \text{ m}^2\right)(L, \text{ m})\,(200 \text{ tubes per m}^2) = 100 \text{ m}$$

or
$$50\pi D^2 L = 100$$

To convert this constrained optimization into an unconstrained optimization, solve for $L$ in the constraint and substitute into the objective function

$$\text{Cost} = 900 + \frac{2200}{\pi} D^{0.5} + \frac{640}{\pi D}$$

The objective function is now in terms of $D$ only, so differentiating and equating to zero yields

$$\frac{d(\text{cost})}{d(D)} = \frac{1100}{\pi D^{0.5}} - \frac{640}{\pi D^2} = 0$$

$$D^* = 0.7 \text{ m}$$

Substituting the optimal value of $D$ back into the constraint gives

$$L^* = 1.3 \text{ m}$$

and the minimum cost is

$$\text{Cost}^* = 900 + (1100)(0.7)^{2.5}(1.3) + (320)(0.7)(1.3)$$

$$= \$1777.45$$

## 8-6 CONSTRAINED OPTIMIZATION USING LAGRANGE MULTIPLIER EQUATIONS

While it is sometimes possible to convert a constrained optimization into an unconstrained one, there are severe limitations to this option. For example, it may not be possible to solve the constraint equations explicitly for variables that are to be substituted into the objective function. The more general and powerful technique is to use the classic Lagrange multiplier equations (8-4) to (8-6).

**Example 8-3** Solve Example 8-2 using the Lagrange multiplier equations.

SOLUTION The statement of the problem is:

| | | |
|---|---|---|
| Minimize | $y = 900 + 1100D^{2.5}L + 320DL$ | (8-13) |
| subject to | $50\pi D^2 L = 100$ | (8-14) |

The two gradient vectors are

$$\nabla y = [(2.5)(1100)D^{1.5}L + 320L]i_1 + (1100D^{2.5} + 320D)i_2$$

and
$$\nabla \phi = 100\pi DL i_1 + 50\pi D^2 i_2$$

In component form the two scalar equations are

$i_1$:
$$2750D^{1.5}L + 320L - \lambda 100\pi DL = 0$$

$i_2$:
$$1100D^{2.5} + 320D - \lambda 50\pi D^2 = 0$$

which along with the constraint

$$50\pi D^2 L = 100$$

provide three simultaneous equations to solve for $D^*$, $L^*$, and $\lambda$. From the $i_1$ equation

$$\lambda = \frac{2750 D^{1.5} + 320}{100\pi D}$$

Substituted into the $i_2$ equation, this yields

$$D^* = 0.7 \text{ m}$$

Substituting this value of $D$ into the constraint gives

$$L^* = 1.3 \text{ m}$$

and

$$\lambda = \frac{(2750)(0.7)^{1.5} + 320}{100\pi(0.7)} = 8.78$$

The minimum cost is

$$y^* = 900 + (1100)(0.7)^{2.5}(1.3) + (320)(0.7)(1.3) = \$1777.45$$

The objective function, Eq. (8-13), is sketched in Fig. 8-4a. So long as either $L$ or $D$ is zero, the cost is \$900. As $L$ and $D$ increase, the cost rises, providing a warped surface that rises as it moves away from the vertical axis. The problem takes on meaning only after introduction of the constraint in Fig. 8-4b, which is a curve in the $LD$ plane. Only the vertical projections from the constraint up to the objective function surface are permitted. The minimum value of the objective function along the projection is the constrained minimum.

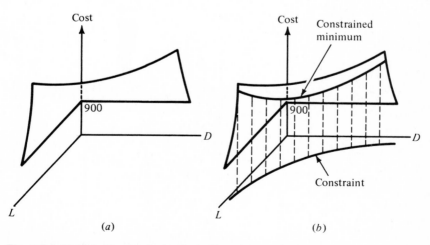

**Figure 8-4** (a) Objective-function surface; (b) with constraint added.

Since Example 8-3 had only one constraint equation, only one $\lambda$ appeared in the Lagrange multiplier equations. There will be the same number of $\lambda$'s as constraints. The execution of the solution requires the solution of a set of equations which are likely to be nonlinear. In Example 8-3 it was possible to solve them by substitution. When this cannot be done, the tools are now available from Chap. 6 to solve the set of nonlinear simultaneous equations by using such techniques as the Newton-Raphson method.

## 8-7 GRADIENT VECTOR IS NORMAL TO THE CONTOUR SURFACE

The purpose of this section and the next is to provide a visualization of the Lagrange multiplier equations. The presentation will not be a proof[†] but a geometric display. The first step in providing the visualization is to show that the gradient vector is normal to the contour line or surface at the point where the gradient is being evaluated. If $y$ is a function of $x_1$ and $x_2$, a contour line on the $x_1$ vs. $x_2$ graph is a line of constant $y$, as shown in Fig. 8-5. From calculus we recall that

$$dy = \frac{\partial y}{\partial x_1} dx_1 + \frac{\partial y}{\partial x_2} dx_2$$

[†]A proof can be found in the book by Wilde and Beightler given in the Additional Readings at the end of the chapter.

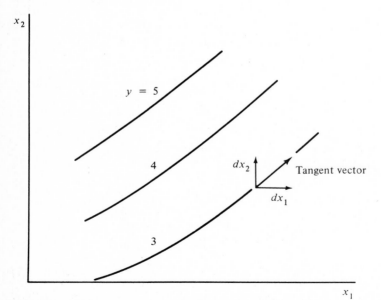

**Figure 8-5** Contour lines (curves of constant $y$) when $y = f(x_1, x_2)$.

Therefore along the line of constant $y$, say the $y = 3$ line,

$$dy = 0 = \frac{\partial y}{\partial x_1} dx_1 + \frac{\partial y}{\partial x_2} dx_2$$

or

$$dx_1 = -dx_2 \frac{\partial y / \partial x_2}{\partial y / \partial x_1} \tag{8-15}$$

Any arbitrary vector in Fig. 8-5 is

$$dx_1 \, \mathbf{i}_1 + dx_2 \, \mathbf{i}_2$$

and any arbitrary unit vector is

$$\frac{dx_1 \, \mathbf{i}_1 + dx_2 \, \mathbf{i}_2}{\sqrt{(dx_1)^2 + (dx_2)^2}} \tag{8-16}$$

The special unit vector $\mathbf{T}$, the one that is tangent to the $y$ = constant line, has $dx_1$ and $dx_2$ related according to Eq. (8-15); substituting $dx_1$ from Eq. (8-15) into Eq. (8-16) yields

$$\mathbf{T} = \frac{-\dfrac{\partial y / \partial x_2}{\partial y / \partial x_1} \mathbf{i}_1 + \mathbf{i}_2}{\sqrt{\left(\dfrac{\partial y / \partial x_2}{\partial y / \partial x_1}\right)^2 + 1}} = \frac{-\dfrac{\partial y}{\partial x_2} \mathbf{i}_1 + \dfrac{\partial y}{\partial x_1} \mathbf{i}_2}{\sqrt{\left(\dfrac{\partial y}{\partial x_2}\right)^2 + \left(\dfrac{\partial y}{\partial x_1}\right)^2}} \tag{8-17}$$

Equation (8-17) is the unit vector that is tangent to the $y$ = const line.

Returning to the gradient vector and dividing by its magnitude to obtain the unit gradient vector $\mathbf{G}$, we have

$$\mathbf{G} = \frac{\nabla y}{|\nabla y|} = \frac{\partial y / \partial x_1 \, \mathbf{i}_1 + \partial y / \partial x_2 \, \mathbf{i}_2}{\sqrt{(\partial y / \partial x_1)^2 + (\partial y / \partial x_2)^2}} \tag{8-18}$$

The relationship between the vectors represented by Eqs. (8-17) and (8-18) is that one is perpendicular to the other. If, as in Fig. 8-6, the components of vector $\mathbf{Z}$ are $b$ and $c$, the perpendicular vector $\mathbf{Z}_1$ has components $c$ and $-b$. Thus, the components are interchanged and the sign of one is reversed.

The important conclusion reached at this point is that since the gradient vector is perpendicular to the tangent vector, *the gradient vector is normal to the line or surface of constant $y$*.

A similar conclusion is applicable in three and more dimensions. In three dimensions, for example, where $y = y(x_1, x_2, x_3)$, the curves of constant value of $y$ become surfaces, as shown in Fig. 8-7. In this case the gradient vector $\nabla y$, which is

$$\nabla y = \frac{\partial y}{\partial x_1} \mathbf{i}_1 + \frac{\partial y}{\partial x_2} \mathbf{i}_2 + \frac{\partial y}{\partial x_3} \mathbf{i}_3 \tag{8-19}$$

is normal to the $y$ = const surface that passes through that point.

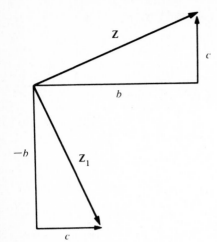

**Figure 8-6** Components of perpendicular vectors.

The magnitude of the gradient vector indicates the rate of change of the dependent variable with respect to the independent variables. Thus, if the surfaces of constant $y$ in Fig. 8-7 are spaced wide apart, the absolute magnitude of the gradient is small. For the time being, however, we are interested only in the fact that the gradient vector is in a *direction* normal to the constant curve or surface.

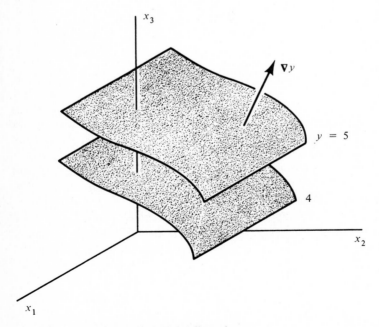

**Figure 8-7** A gradient vector in three dimensions.

## 8-8 VISUALIZATION OF THE LAGRANGE MULTIPLIER METHOD IN TWO DIMENSIONS

An optimization of a problem with two variables of optimization can be displayed graphically to help clarify the operation of the Lagrange multiplier equations. Suppose that the function $y$ is to be minimized, where

$$y = 2x_1 + 3x_2 \qquad (8\text{-}20)$$

subject to the constraint

$$x_1 x_2^2 = 48 \qquad (8\text{-}21)$$

The Lagrange multiplier equations are

$$(2 - \lambda x_2^2)\mathbf{i}_1 + (3 - 2\lambda x_1 x_2)\mathbf{i}_2 = 0 \quad \text{and} \quad x_1 x_2^2 = 48$$

which when solved yield the solution $x_1^* = 3$, $x_2^* = 4$, and $y^* = 18$. Some lines of constant $y$ and the constraint equation are shown in Fig. 8-8. With such a graph available it is possible almost instinctively to locate the $(x_1 x_2)$ position that provides the constrained optimum, point $A$ in Fig. 8-8. The process of visually arriving at this solution might be one of following along the constraint line until the constraint and the line of constant $y$ are "parallel." Stated more precisely, the tangent vectors to the curves have the same direction.

A more convenient means of requiring that the tangent vectors have the same direction is to require that the normal vectors to the curves have the same direction. Since the gradient vector is normal to a contour line, the mathematical

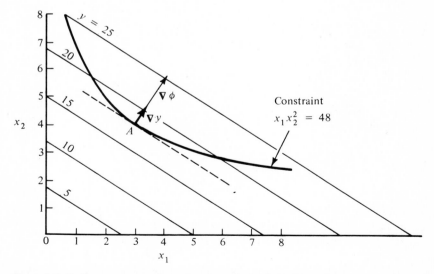

**Figure 8-8** Optimum occurs where the constraint and the lines of constant $y$ have a common normal.

statement of this requirement is

$$\nabla y - \lambda \nabla \phi = 0 \qquad (8\text{-}22)$$

The introduction of a constant, $\lambda$, is necessary because the magnitudes of $\nabla y$ and $\nabla \phi$ may be different and the vectors may even be pointing in opposite directions. The only requirement is that the two gradient vectors be collinear. The constraint equation is satisfied because the only points being considered are those along the constraint equation.

## 8-9 TEST FOR MAXIMUM OR MINIMUM

In mathematics books that treat optimization, test procedures for determining the nature of the critical point receive a major emphasis. These test procedures decide whether the point is a maximum, minimum, saddle point, ridge, or valley. These procedures are generally not so important in physical problems, which are our concern, because the engineer usually has an insight into whether the result is a maximum or minimum. It also seems that the occurrence of saddle points, ridges, and valleys is rare in physical problems. In addition, the tests based on pure calculus usually become prohibitive when dealing with more than about three variables anyway, and we are especially interested in systems with large numbers of components. If a test is made, it usually consists of testing points in the neighborhood of the optimum.

On the other hand, a brief discussion of the classical tests for maximum and minimum provide further insight into the nature of the optimization process. The discussion will be limited to the optimization of one and two variables of an unconstrained function.

Consider first the case of one independent variable $y = y(x)$ for which the minimum is sought. Suppose that the point $x = a_1$ is the position at which the minimum is expected to occur. To test whether the value of $y$ at this point $y(a_1)$ is truly a minimum, move slightly in both possible directions from $x = a_1$ to see if a lower value of $y$ can be found. If a lower value is available, $y(a_1)$ is not the minimal value. For the mathematical check, expand $y$ in a Taylor series (see Sec. 6-9) about the point $x = a_1$

$$y(x) = y(a_1) + \frac{dy}{dx}(x - a_1) + \frac{1}{2}\frac{d^2 y}{dx^2}(x - a_1)^2 + \cdots \qquad (8\text{-}23)$$

First examine moves so small that the $(x - a_1)^2$ terms and higher-order terms can be ignored; if $dy/dx > 0$, then $y(x) > y(a_1)$ for a value of $x > a_1$ and $y(a_1)$ is still the acknowledged minimum. When $x$ moves to a value less than $a_1$, however, $y(x) < y(a_1)$ and $y(a_1)$ will not be a minimum. In a similar manner, it can be shown that when $dy/dx < 0$, a lower value than $y(a_1)$ can be found for $y(x)$. The only solution to the dilemma is for $dy/dx$ to equal zero, which is the classical requirement for the optimum.

Including the influence of the next term of Eq. (8-23), $(\frac{1}{2})(d^2y/dx^2)(x - a_1)^2$, we observe that a move of $x$ in either direction from $a_1$ results in a positive value of $(x - a_1)^2$, so that the sign of the second derivative decides whether the optimum is a maximum or minimum.

Minimum:
$$\frac{dy}{dx} = 0 \quad \text{and} \quad \frac{d^2y}{dx^2} > 0$$

Maximum:
$$\frac{dy}{dx} = 0 \quad \text{and} \quad \frac{d^2y}{dx^2} < 0$$

The foregoing line of reasoning will now be extended to a function of two variables $y(x_1, x_2)$. Suppose that the expected minimum occurs at the point $(a_1, a_2)$ and that to verify this position the point is shifted infinitesimally away from $(a_1, a_2)$ in all possible directions. The Taylor expansion for a function of two variables is

$$y(x_1, x_2) = y(a_1, a_2) + y_1'(x_1 - a_1) + y_2'(x_2 - a_2) + (\tfrac{1}{2})y_{11}''(x_1 - a_1)^2$$
$$+ y_{12}''(x_1 - a_1)(x_2 - a_2) + (\tfrac{1}{2})y_{22}''(x_2 - a_2)^2 + \cdots \quad (8\text{-}24)$$

where the prime on the $y$ refers to a partial differentiation with respect to the subscript.

When $x_1$ and $x_2$ move slightly off $(a_1, a_2)$, both of the first derivatives $y_1'$ and $y_2'$ must be zero in order to avoid some position where $y(x_1, x_2) < y(a_1, a_2)$.

The second-order terms decide whether the optimum is a maximum or minimum. If the combination is always positive regardless of the signs of $x_1 - a_1$ and $x_2 - a_2$, the optimum is a minimum. If the combination is always negative, the optimum is a maximum.

The test for maximum and minimum is as follows. The second derivatives are structured in matrix form and the value of the determinant is called $D$, where

$$D = \begin{vmatrix} y_{11}'' & y_{12}'' \\ y_{21}'' & y_{22}'' \end{vmatrix}$$

Then if $D > 0$ and $y_{11}'' > 0$, the optimum is a minimum. If $D > 0$ and $y_{11}'' < 0$, the optimum is a maximum.

**Example 8-4** Determine the optimal values of $x_1$ and $x_2$ for the function

$$y = \frac{x_1^2}{4} + \frac{2}{x_1 x_2} + 4x_2$$

and test whether the point is a maximum or minimum.

SOLUTION The first derivatives are

$$\frac{\partial y}{\partial x_1} = \frac{x_1}{2} - \frac{2}{x_1^2 x_2} \quad \text{and} \quad \frac{\partial y}{\partial x_2} = -\frac{2}{x_1 x_2^2} + 4$$

Equating these derivatives to zero and solving simultaneously yields

$$x_1^* = 2 \quad \text{and} \quad x_2^* = \tfrac{1}{2}$$

The second derivatives evaluated at $x_1^*$ and $x_2^*$ are

$$\frac{\partial^2 y}{\partial x_1^2} = \frac{3}{2} \qquad \frac{\partial^2 y}{\partial x_2^2} = 16 \qquad \frac{\partial^2 y}{\partial x_1 \, \partial x_2} = 2$$

The determinant

$$\begin{vmatrix} \frac{3}{2} & 2 \\ 2 & 16 \end{vmatrix} > 0 \quad \text{and} \quad y_{11}'' > 0$$

so this optimal point is a minimum.

## 8-10 SENSITIVITY COEFFICIENTS

There is often an additional valuable step beyond determination of the optimal value of the objective function and the state point at which the optimum occurs. After the optimum is found, subject to one or more constraints, the question that logically arises is: What would be the effect on the optimal value of slightly relaxing the constraint? In a physical situation this question occurs, for example, in analyzing how much the capacity of a system could be increased by enlarging one of the components whose performance characteristic is one of the constraint equations.

In Example 8-3, where the cost of the heat exchanger

$$\text{Cost} = 900 + 1100 D^{2.5} L + 320 DL$$

was minimized subject to the constraint

$$50\pi D^2 L = 100$$

the question might be phrased: What would be the increase in minimum cost, cost*, if 101 linear meters of tubes were required rather than the original 100? To analyze this particular example, replace the specific value of 100 by a general symbol $H$ and perform the optimization by the method of Lagrange multipliers. The result would be

$$D^* = 0.7 \text{ m} \qquad L^* = 0.013H$$

$$\text{Cost}^* = 900 + (1100)(0.7)^{2.5}(0.013H) + (320)(0.7)(0.013H)$$

$$= 900 + 8.78H$$

We are interested in the variation of cost* with $H$ or, more specifically, a term called the *sensitivity coefficient* (SC), which is

$$\text{SC} = \frac{\partial(\text{cost}^*)}{\partial H}$$

In this example, SC = 8.78; thus at the optimal proportions an extra meter of tube for the heat exchanger would cost $8.78 more than the original cost. Referring back to the solution of Example 8-3, we note the remarkable fact that the sensitivity coefficient is precisely equal to the Lagrange multiplier $\lambda$. This equality of the SC to $\lambda$ is true not only for this particular example but in general. Also, if there is more than one constraint, the various sensitivity coefficients are equal to the corresponding Lagrange multipliers, $SC_1 = \lambda_1, \ldots, SC_n = \lambda_n$.

The optimization process by Lagrange multipliers therefore offers an additional piece of useful information for possible adjustment of the physical system following preliminary optimization.

## 8-11 INEQUALITY CONSTRAINTS

The method of Lagrange multipliers applies only to the situations where the constraints are equalities, and the method cannot be used directly with inequality constraints. This limitation should not completely rule out the employment of the method when inequalities arise, because a combination of intuition and several passes at the problem with the method may yield a solution.

As an example, suppose that the capacity of a system is said to be equal to or greater than 150 kW. It would be a rare case when a lower-cost system would provide 160 kW compared with the one providing 150 kW. This constraint would almost certainly be used as an equality constraint of 150 kW. As a further example, suppose that the temperature at some point in the process must be equal to or less than 320°C. In the first attempt at the problem, ignore the temperature constraint and after the optimal conditions are determined check to see if the temperature in question is above 320°C. If it is not, the constraint is not effective. If the temperature is above 320°C, rework the problem with the equality constraint of 320°C.

## PROBLEMS

**8-1** For the function

$$y = \frac{1}{4x_1} + 4x_1^2 x_2 + \frac{1}{x_2}$$

(*a*) Determine the positive optimum values of $x_1$ and $x_2$.

(*b*) Use the techniques of Sec. 8-9 to test the solution to determine whether the optimum is a maximum or minimum.

Ans.: (*a*) $x_1^* = \frac{1}{4}, x_2^* = 2$.

**8-2** The dimensions of a rectangular duct are to be chosen so that the duct of maximum cross-sectional area can be placed, as shown in Fig. 8-9, in the opening in a bar joist.

(*a*) Set up the objective function and constraint in terms of $h$ and $w$.

(*b*) Using the method of Lagrange multipliers for constrained optimization, determine the optimal values of $h$ and $w$.

Ans.: $h^* = 0.3$ m.

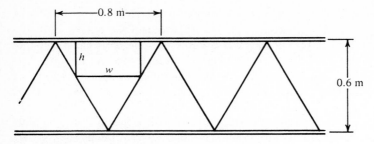

**Figure 8-9** Duct in bar joist in Prob. 8-2.

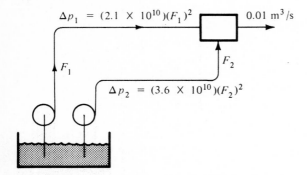

**Figure 8-10** Parallel pump-pipe assemblies in Prob. 8-3.

**8-3** Two parallel pump-pipe assemblies shown in Fig. 8-10 deliver water from a common source to a common destination. The total volume flow rate required at the destination is $0.01$ m$^3$/s. The drops in pressure in the two lines are functions of the square of the flow rates,

$$\Delta p_1, \text{Pa} = 2.1 \times 10^{10} F_1^2 \quad \text{and} \quad \Delta p_2, \text{Pa} = 3.6 \times 10^{10} F_2^2$$

where $F_1$ and $F_2$ are the respective flow rates in cubic meters per second. The two pumps have the same efficiency, and the two motors that drive the pumps also have the same efficiency.

(a) It is desired to minimize the total power requirement. Set up the objective function and constraints in terms of $F_1$ and $F_2$.

(b) Solve for the optimal values of the flow rates that result in minimum total water power using the method of Lagrange multipliers.

**Ans.:** $F_1 = 0.00567$.

**8-4** A steel framework, as in Fig. 8-11, is to be constructed at a minimum cost. The cost in dollars of all the horizontal members in one orientation is $200x_1$ and in the other horizontal orientation $300x_2$. The cost in dollars of all vertical members is $500x_3$. The frame must enclose a total volume of $900$ m$^3$.

(a) Set up the objective function for total cost and the constraint(s) in terms of $x_1$, $x_2$, and $x_3$.

(b) Using the method of Lagrange multipliers for constrained optimization, determine the optimal values of the dimensions and the minimum cost.

**Ans.:** Minimum cost = \$9000.

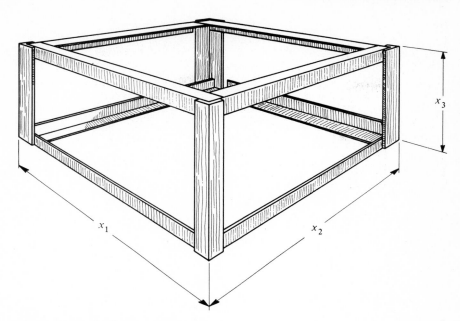

**Figure 8-11** Steel framework in Prob. 8-4.

**8-5** A flow rate of 15 m³/s of gas at a temperature of 50°C and a pressure of 175 kPa is to be compressed to a final pressure of 17,500 kPa. The choice of compressor type is influenced by the fact that centrifugal compressors can handle high-volume flow rates but develop only low pressure ratios per stage. The reciprocating compressor, on the other hand, is suited to low-volume flow rates but can develop high pressure ratios. To combine the advantages of each, the compression will be carried out by a centrifugal compressor in series with a reciprocating compressor, as shown in Fig. 8-12.

The intercooler returns the temperature of the gas to 50°C. Assume that the gas obeys perfect-gas laws. The equations for the costs of the compressors are

$$C_c = 70Q_0 + 1600\,\frac{p_1}{p_0} \qquad C_r = 200Q_1 + 800\,\frac{p_2}{p_1}$$

where $C_c$ = first cost of centrifugal compressor, dollars
$C_r$ = first cost of reciprocating compressor, dollars
$Q$ = volume flow rate, m³/s

(*a*) Set up the objective function for the total first cost and the constraint equation in terms of the pressure ratios.

(*b*) Using the method of Lagrange multipliers for constrained optimization, solve for the optimal pressure ratios and minimum total cost.

**Ans.:** Minimum cost = \$24,100.

**8-6** A solar collector and storage tank, shown in Fig. 8-13*a*, is to be optimized to achieve minimum first cost. During the day the temperature of water in the storage vessel is elevated from 25°C (the minimum useful temperature) to $t_{max}$, as shown in Fig. 8-13*b*. The collector receives 260 W/m² of solar energy, but there is heat loss from the collector to ambient air by convection. The convection coefficient is 2 W/(m² · K), and the average tem-

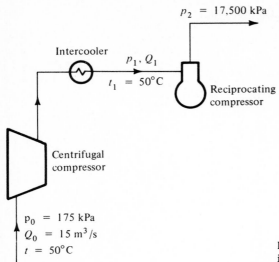

$p_2 = 17,500$ kPa

Intercooler

$p_1, Q_1$

$t_1 = 50°C$

Reciprocating compressor

Centrifugal compressor

$p_0 = 175$ kPa
$Q_0 = 15$ m³/s
$t = 50°C$

**Figure 8-12** Staged compression in Prob. 8-5.

perature difference during the 10-hour day is $(25 + t_{max})/2$ minus the ambient temperature of 10°C.

The energy above the minimum useful temperature of 25°C that is to be stored in the vessel during the day is 200,000 kJ. The density of water is 1000 kg/m³, and its specific heat is 4.19 kJ/(kg · K). The cost of the solar collector in dollars is $20A$, where $A$ is the area in square meters, and the cost of the storage vessel in dollars is $101.5V$, where $V$ is the volume in cubic meters.

(a) Using $A$ and $V$ as the variables, set up the objective function and constraint to optimize the first cost.

(b) Develop the Lagrange multiplier equations and verify that they are satisfied by $V = 1.2$ m³ and $A = 29.2$ m².

**Ans.:** (a) $y = 20A + 101.5V$, subject to $A(230 - 47.7/V) = 5555$.

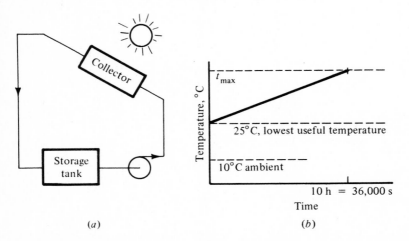

$t_{max}$

25°C, lowest useful temperature

10°C ambient

10 h = 36,000 s

Time

Temperature, °C

Collector

Storage tank

(a)                    (b)

**Figure 8-13** (a) Collector and storage tank and (b) temperature variation in Prob. 8-6.

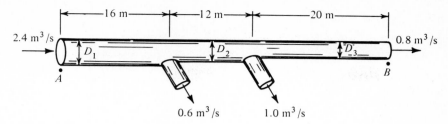

**Figure 8-14** Air duct in Prob. 8-7.

**8-7** Determine the diameters of the circular air duct in the duct system shown schematically in Fig. 8-14 so that the drop in static pressure between points $A$ and $B$ will be a minimum.

## Further information

Quantity of sheet metal available for the system, 60 $m^2$.
Pressure drop in a section of straight duct, $f(L/D)(V^2/2)\rho$.
Use a constant friction factor $f = 0.02$.
Air density $\rho$, 1.2 $kg/m^3$.
Neglect the influence of changes in velocity pressure.
Neglect the pressure drop in the straight-through section past an outlet.

    (*a*) Set up the objective function and constraint(s) in terms of $D_1$, $D_2$, and $D_3$.
    (*b*) Using the method of Lagrange multipliers for constrained optimization, solve for the optimal values of the diameters.
    **Ans.:** 0.468, 0.425, and 0.325 m.

**8-8** Load dispatchers for electric utilities attempt to operate the most efficient combination of generating units that satisfies the total load requirement. If the input-output curves for two units are as shown in Fig. 8-15, where $O_1$ and $O_2$ are the outputs of units 1 and 2, respectively, and $I_1$ and $I_2$ are the inputs, respectively, and where the total load is $T$, use the method of Lagrange multipliers to prove that the optimum loading $O_1^*$ and $O_2^*$ occur when the slopes of the curves are equal.

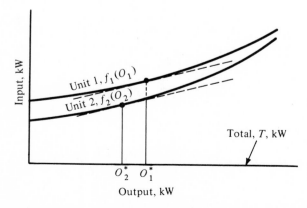

**Figure 8-15** Optimal loading of two units in a power generating station, Prob. 8-8.

**8-9** A cylindrical oil-storage tank is to be constructed for which the following costs apply:

|  | Cost per square meter |
| --- | --- |
| Metal for sides | $30.00 |
| Combined costs of concrete base and metal bottom | 37.50 |
| Top | 7.50 |

The tank is to be constructed with dimensions such that the cost is minimum for whatever capacity is selected.

(*a*) One possible approach to selecting the capacity is to build the tank large enough for an additional cubic meter of capacity to cost $8. (Note that this does not mean $8 per cubic meter average for the entire tank.) What is the optimal diameter and optimal height of the tank?

Ans.: 15 m, 11.25 m.

(*b*) Instead of the approach used in part (*a*), the tank is to be of such a size that the cost will be $9 per cubic meter average for the entire storage capacity of the tank. Set up the Lagrange multiplier equations and verify that they are satisfied by an optimal diameter of 20 m and an optimal height of 15 m.

**8-10** A rectangular duct mounted beneath the beams in a building, as shown in Fig. 8-16, is to have a cross-sectional area of 0.8 m$^2$. The cost of the duct for the required length is $150 per meter of perimeter. The building must be heightened by the amount of the height of the duct, and that cost is $0.80 per millimeter. After the duct has been sized to provide minimum total cost, the possibility of enlarging the duct is explored. What is the additional cost per square meter of cross-sectional area of a very small increase in the area?

Ans.: $642 per square meter.

**8-11** An electric-power generating and distribution system consists of two generating plants and three loads, as shown in Fig. 8-17. The loads are as follows: load $A$ = 40 MW, load $B$ = 60 MW, and load $C$ = 30 MW. The losses in the lines are given in Table 8-1, where the loss in line $i$ is a function of the power $p_i$ in megawatts carried by the line. To be precise, the line loss should be specified as a function of the power at a certain point in the line, e.g., the entrance or exit, but since the loss will be small relative to the power carried, use $p_i$ at the point in the line most convenient for calculation.

As a first approximation in the load balances, assume that $p_5$ leaving load $A$ equals $p_5$ entering load $B$, and recalculate, if necessary, after the first complete solution.

Assuming that the two generating plants are equally efficient, use the method of Lagrange multipliers to compute the optimum amount of power to be carried by each of the lines for the most efficient operation.

Ans.: 24.3, 20.3, 40, 46.4, 4.6, 31 MW.

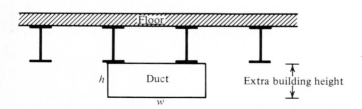

**Figure 8-16** Duct mounted under beams in Prob. 8-10.

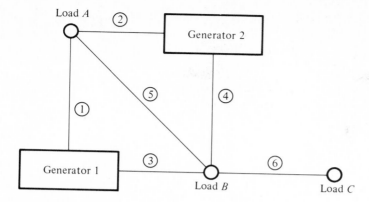

**Figure 8-17** Generating and distribution system in Prob. 8-11.

**Table 8-1 Line losses in Prob. 8-11**

| Line | Loss, MW |
|------|----------|
| 1 | $0.0010p_1^2$ |
| 2 | $0.0012p_2^2$ |
| 3 | $0.0007p_3^2$ |
| 4 | $0.0006p_4^2$ |
| 5 | $0.0008p_5^2$ |
| 6 | $0.0011p_6^2$ |

**8-12** The power-distribution system shown in Fig. 8-18 has a source voltage of 220 V at point 1 and must supply power to positions 3 and 4 at 210 and 215 V, respectively, with a current of 200 and 300 A, respectively. The electrical resistance $R$ $\Omega$ is a function of the area and length of the conductor: $R, \Omega = 17.2 \times 10^{-9}L/A$, where $L$ is the length of conductor in meters and $A$ is the area of conductor in square meters.

(a) Set up the objective function for the total volume of conductor and the constraint(s) in terms of $A_{1-2}, A_{2-4}$, and $A_{2-3}$.

(b) Verify, using all the Lagrange multiplier equations, that $A_{1-2}^* = 0.00273$ m$^2$.

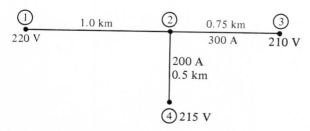

**Figure 8-18** Power-distribution network in Prob. 8-12.

## ADDITIONAL READINGS

Bowman, F., and F. A. Gerard: *Higher Calculus*, Cambridge University Press, London, 1967.

Brand, L.: *Advanced Calculus*, Wiley, New York, 1955.

Kaplan, W.: *Advanced Calculus*, Addison-Wesley, Reading, Mass., 1952.

Taylor, A. E.: *Advanced Calculus*, Ginn, Boston, Mass., 1955.

Wilde, D. J., and C. S. Beightler: *Foundations of Optimization*, Prentice-Hall, Englewood Cliffs, N.J., 1967.

# NINE

## SEARCH METHODS

## 9-1 OVERVIEW OF SEARCH METHODS

The name *search method* should not suggest a helter-skelter wandering through the variables of optimization until a favorable value of the objective function, subject to the constraints, has been found. Instead, search methods generally fall into two categories, *elimination* and *hill-climbing techniques*. In both there is a progressive improvement throughout the course of the search.

In one respect search methods are a paradox. They are often the ultimate approach if other methods of optimization (such as Lagrange multipliers, dynamic programming, etc.) fail. From another standpoint search methods are unsatisfying because there is no one systematic procedure that is followed and the literature is laden with different variations of search techniques. When a problem is found in which the standard techniques experience difficulty, another variation is developed. In the short treatment of this chapter we shall adhere to several approaches that are quite classical, have broad applicability, and flow naturally from the calculus principles presented in Chap. 8.

The outline of the coverage is as follows:

1. Single variable
   *a.* Exhaustive
   *b.* Efficient
       i. Dichotomous
       ii. Fibonacci

2. Multivariable, unconstrained
   a. Lattice
   b. Univariate
   c. Steepest ascent
3. Multivariable, constrained
   a. Penalty functions
   b. Search along a constraint

In the calculus method of optimization presented in Chap. 8, calculating the numerical value of the objective function was virtually the last step in the process. The major effort in the optimization was determining the values of the independent variables that provide the optimum. In optimization by means of search methods, an opposite sequence is followed in that values of the objective function are determined and conclusions are drawn from the values of the function at various combinations of independent variables.

## 9-2 INTERVAL OF UNCERTAINTY

An accepted feature of search methods is that the precise point at which the optimum occurs will never be known, and the best that can be achieved is to specify the *interval of uncertainty*. This is the range of the independent variable(s) in which the optimum is known to exist. An interval of uncertainty prevails because the search method computes the value of the function only at discrete values of the independent variables.

## 9-3 EXHAUSTIVE SEARCH

Of the various search methods used in single-variable problems, the exhaustive search is the least imaginative but most widely used, and justifiably so. The method consists of calculating the value of the objective function at values of $x$ that are spaced uniformly throughout the interval of interest. The interval of interest $I_0$ (Fig. 9-1) is divided here into eight equal intervals. Assume that the values of $y$ are calculated at the seven positions shown. In this example the maximum lies between $x_A$ and $x_B$, so the final interval of uncertainty $I$ is

$$I = \frac{2I_0}{8} = \frac{I_0}{4}$$

If two observations are made, the final interval of uncertainty is $2I_0/3$; if three observations are made, the final interval of uncertainty is $2I_0/4$, and in general

$$\text{Final interval of uncertainty} = I = \frac{2I_0}{n+1} \tag{9-1}$$

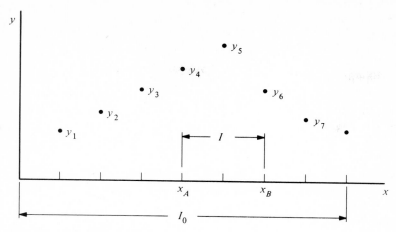

Figure 9-1 Exhaustive search.

## 9-4 UNIMODAL FUNCTIONS

The dichotomous and Fibonacci methods, introduced next, are applicable to *unimodal* functions. A unimodal function is one having only one peak (or valley) in the interval of interest. Figure 9-2 shows several unimodal functions. The dichotomous and Fibonacci methods can successfully handle not only the smooth curve like Fig. 9-2*a* but also the nondifferentiable function like Fig. 9-2*b* or even a discontinuous function like Fig. 9-2*c*.

In optimizing nonunimodal functions where there are several peaks (or valleys), the function can be subdivided into several parts and each part processed separately as a unimodal function.

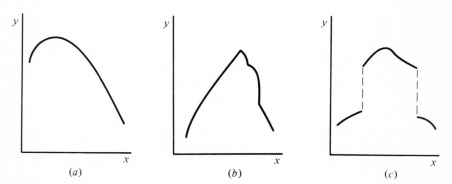

Figure 9-2 Unimodal functions.

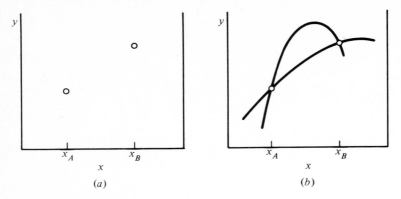

Figure 9-3 Two test points on a unimodal function.

## 9-5 ELIMINATING A SECTION BASED ON TWO TESTS

Knowledge of the value of the objective function at two different positions in the interval of interest is sufficient to eliminate a portion of the region of a unimodal function. Suppose that a maximum value of $y$ is sought in the function shown in Fig. 9-3$a$. The magnitude of $y$ is known at two values of $x$, designated $x_A$ and $x_B$. From this information it is possible to eliminate the region to the left of $x_A$. The region to the right of $x_A$ must still be retained. It cannot be determined whether the maximum lies between $x_A$ and $x_B$ or to the right of $x_B$ because the maximum could reside in either interval, as shown in Fig. 9-3$b$.

## 9-6 DICHOTOMOUS SEARCH

The concept of the *dichotomous search* follows closely the discussion of placing two test points and asks: Where should the two points be placed in the interval of uncertainty in order to eliminate the largest possible region? A little reflection will show that placing the points as near the center as possible while maintaining distinguishability of the $y$ values will result in elimination of almost half the original interval of uncertainty. Figure 9-4 shows the two points placed symmetric to the center with a spacing of $\epsilon$ between. With the values of $y$ as shown, the region to the left of $x_A$ can be eliminated, so the remaining interval is $(I_0 + \epsilon)/2$.

The next pair of observations is made in a similar manner in the remaining interval of uncertainty, resulting in the further reduction of the interval by nearly one-half. In general, the remaining interval of uncertainty $I$ is

$$I = \frac{I_0}{2^{n/2}} + \epsilon \left(1 - \frac{1}{2^{n/2}}\right) \tag{9-2}$$

where $n$ is the number of tests $(2, 4, 6, \text{etc.})$.

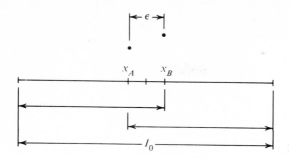

**Figure 9-4** Dichotomous search.

## 9-7 FIBONACCI SEARCH

The most efficient of the single-variable search techniques is the Fibonacci method. This method was first presented by Kiefer,[1] who applied the Fibonacci number series, which was named after the thirteenth-century mathematician. The rule for determining a Fibonacci number $F$ is

$$F_0 = 1$$

$$F_1 = 1$$

$$F_i = F_{i-2} + F_{i-1} \qquad \text{for } i \geqslant 2$$

Thus, after the first two Fibonacci numbers are available, each number thereafter is found by summing the two preceding numbers. The Fibonacci series starting with the index zero is therefore 1, 1, 2, 3, 5, 8, 13, 21, 34, 55, $F_{10} = 89$, etc.

The steps in executing a Fibonacci search are as follows:

*Step 1.* Decide how many observations will be made and call this number $n$.

*Step 2.* Place the first observation in $I_0$ so that the distance from one end is $I_0(F_{n-1}/F_n)$.

*Step 3.* Place the next observation in the interval of uncertainty at a position that is symmetric to the existing observation. According to the relative values of these observations, eliminate either the region to the right of the right point or to the left of the left point. Continue placing a point and eliminating a region until one point remains to be placed. At this stage there will be one observation directly in the center of the interval of uncertainty.

*Step 4.* Place the last observation as close as possible to this center point and eliminate half the interval.

**Example 9-1** Perform a Fibonacci search to find the maximum of the function $y = -(x)^2 + 4x + 2$ in the interval $0 \leqslant x \leqslant 5$.

SOLUTION

*Step 1.* Arbitrarily choose $n = 4$.

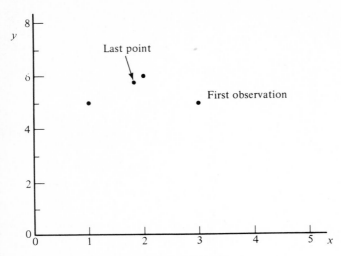

**Figure 9-5** Fibonacci search in Example 9-1.

*Step 2.* Place the first observation a distance $I_0(F_3/F_4)$ from the left end, as in Fig. 9-5. This distance is $\frac{3}{5}I_0$ or $(\frac{3}{5})(5)$. The current interval of uncertainty is 0 to 5 with an observation at 3.

*Step 3.* The next observation symmetric in the interval of uncertainty to 3 locates this observation at $x = 2$. By making use of the relative values of $y$ at $x = 2$ and $x = 3$, the section $3 < x \leqslant 5$ can be eliminated. The interval of uncertainty is now $0 \leqslant x \leqslant 3$, with the observation at $x = 2$ available. Placing the third point symmetric to the $x = 2$ observation locates it at $x = 1$. The relative values of $y$ at $x = 1$ and $x = 2$ permit elimination of $0 \leqslant x < 1$.

*Step 4.* The final point is positioned as close as possible to the $x = 2$ point, which is now in the center of the existing interval of uncertainty. If this final point is placed at $x = 2 - \epsilon$, the final interval of uncertainty is $2 - \epsilon \leqslant x \leqslant 3$.

The final interval of uncertainty in Example 9-1 was $I_0/5 + \epsilon$. In general, if $n$ observations are made, the final interval of uncertainty is $I_0/F_n + \epsilon$.

## 9-8 COMPARATIVE EFFECTIVENESS OF SEARCH METHODS

A measure of the efficiency of a search method is the reduction ratio (RR), proposed by Wilde,[2] which is defined as the ratio of the original interval of uncer-

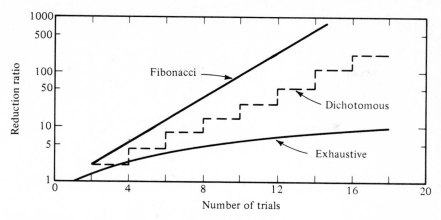

**Figure 9-6** Comparison of reduction ratios of search methods.

tainty to the interval after $n$ trials $I_n$

$$RR = \frac{I_0}{I_n} \qquad (9\text{-}3)$$

If the penalty on the reduction ratio imposed by $\epsilon$ in the dichotomous and Fibonacci methods is neglected, the reduction ratios are

$$RR = \begin{cases} \dfrac{n+1}{2} & \text{exhaustive} & (9\text{-}4) \\[2mm] 2^{n/2} & \text{dichotomous} & (9\text{-}5) \\[2mm] F_n & \text{Fibonacci} & (9\text{-}6) \end{cases}$$

The RR of the Fibonacci method, for example, is 377 after 13 trials. The comparison is shown graphically in Fig. 9-6.

## 9-9 ASSESSMENT OF SINGLE-VARIABLE SEARCHES

The efficiency of such methods as the dichotomous and Fibonacci searches compared with the exhaustive search is impressive. Realistically, however, the single-variable optimizations are not the ones where high efficiencies are needed. In most single-variable optimizations in thermal systems, if the search is performed on a computer, the exhaustive search requires only a few more dollars of computer time and is easier to program. Only if the calculations required for each point are extremely lengthy is there any significant advantage in one of the efficient search methods, and such extensive calculations for single-variable optimizations are rare.

Another potential use of efficient single-variable searches is as a step in a multivariable search. An additional restriction arises, however, and that is the risk of the function being nonunimodal. Later in this chapter a single-variable search will be shown (Fig. 9-11, for example) making a cut across some contours in a multivariable problem. If those contours are kidney-shaped, the single-variable cut may describe a nonunimodal function to which the dichotomous and Fibonacci methods do not apply.

We move now to multivariable problems, where exhaustive searches may be prohibitively expensive and efficient techniques are demanded.

## 9-10 MULTIVARIABLE, UNCONSTRAINED OPTIMIZATION

High efficiency in multivariable searches may be crucial. The most significant optimizations in thermal systems involve many components and thus many variables. Furthermore, the complexity of the equations of a many-component system makes the algebra extremely tedious if a calculus method is used. This complexity breeds errors in formulation. In many multivariable situations, the number of calculations using an exhaustive search introduces considerations of computer time. Suppose, for example, that the optimum temperatures are sought for a seven-stage heat-exchanger chain. Assume that 10 different outlet temperatures for each heat exchanger will be investigated in all combinations with the outlet temperatures of the other heat exchangers. The total number of combinations explored will be $10^7$, which, if the calculation is at all complex makes computer time a definite concern.

The single-variable dichotomous and Fibonacci methods are elimination methods, because after each new point in the Fibonacci search (or pair of points in the dichotomous search) a portion of the interval of uncertainty is eliminated. The three methods for optimization of an unconstrained multivariable problem, to be explained next, are *hill-climbing techniques*, because the calculation is always moving in a way that improves the objective function. The three methods are (1) lattice search, (2) univariate search, and (3) steepest ascent.

## 9-11 LATTICE SEARCH

The procedure in the lattice search is to start at one point in the region of interest and check a number of points in a grid surrounding the central point. The surrounding point having the largest value (if a maximum is being sought) is chosen as the central point for the next search. If no surrounding point provides a greater value of the function than the central point, the central point is the maximum. A frequent practice is first to use a coarse grid and after the maximum has been found for that grid subdivide the grid into smaller elements for a further search, starting from the maximum of the coarse grid.

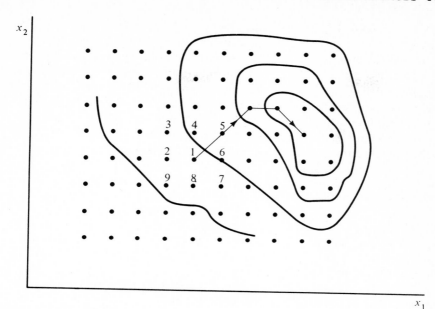

**Figure 9-7** A lattice search.

As an example of the progression to the maximum of a function of two variables, a grid is superimposed over the contour lines of a function in Fig. 9-7. The starting point can be selected near the center of the region unless there is some advance knowledge of where the maximum exists. Call the first point 1 and evaluate the function at points 1 to 9. In this case, the maximum value of the function occurs at point 5, so point 5 becomes the central point for the next search.

Our special objective in applying the lattice search is to improve the efficiency compared with an exhaustive search. Since efficiency is a goal, it is attractive to consider saving information that has already been calculated and can be used again. For example, in Fig. 9-7 after moving from point 1 to point 5, the values of the function at points 1, 4, 5, and 6 are needed in deciding which way to move from point 5. Writing a computer program to accomplish this task is tricky, however, particularly if the number of independent variables exceeds the number of dimensions possible for subscripted variables on the computer being used. It is therefore likely that the functions will be evaluated at all nine points, in the case of two variables, for each central-point location.

No definite statement can be made about the comparative efficiency of the lattice search and the exhaustive search because the nature of the function dictates the number of trials required in the lattice search. In general, however, the ratio of the lattice trials to exhaustive trials decreases as the number of variables increases and the grid becomes finer.

## 9-12  UNIVARIATE SEARCH

In the univariate search, the function is optimized with respect to one variable at a time. The starting procedure is to substitute trial values of all but one independent variable in the function and optimize the resulting function in terms of the one remaining variable. That optimal value is then substituted into the function and the function optimized with respect to another variable. The function is optimized with respect to each variable in sequence with the optimal value of a variable substituted into the function for the optimization of the succeeding variables. The process is shown graphically in Fig. 9-8 for a function of two variables. Along the line of constant $x_1$, which is the initial choice, the value of $x_2$ giving the optimal value of $y$ is determined. This position is designated as point 1. With the value of $x_2$ at point 1 substituted into the function, the function is optimized with respect to $x_1$, which gives point 2. The process continues until the successive change of the dependent or independent variables is less than a specified tolerance.

The method chosen for performing the single-variable optimization may be to use calculus, where the task becomes one of solving one equation (usually a nonlinear one) for one variable. It is also possible to use a single-variable search, exhaustive or efficient, e.g., Fibonacci or dichotomous. It is this use of the efficient single-variable search that is probably the most significant application of this type of search method in thermal-system optimization.

The univariate search can fail when a ridge occurs in the objective function. In the function in Fig. 9-9, for example, if a trial value of $x_1$ is selected as shown, the optimal value of $x_2$ lies on the ridge. When this optimal value of $x_2$ is substituted, the attempt to optimize with respect to $x_1$ does not dislodge the point from the ridge even though the optimum has not been reached.

From the purely mathematical point of view, getting hung up on the ridge would appear to be a serious deficiency of the univariate search. In physical sys-

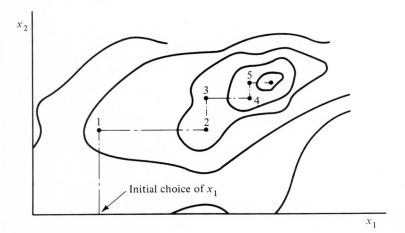

Figure 9-8  Univariate search.

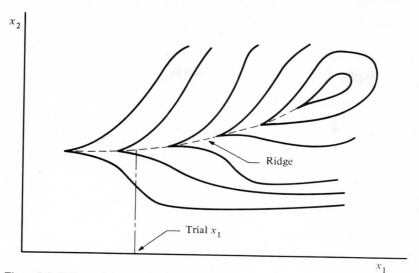

**Figure 9-9** Failure of the univariate search at a ridge.

tems, however, the occurrence of ridges is rare because nature avoids discontinuities of both functions and derivatives of functions. Caution is needed, however, because even though a true ridge does not exist, problems may arise if the contours are very steep. If the interval chosen for the univariate search is too large, the process may stop at a nonoptimal point, such as point $A$ in Fig. 9-10.

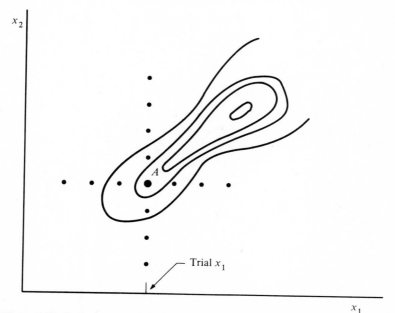

**Figure 9-10** Erroneous conclusion because the interval of search is too large.

Another note of caution is that even though ridges may not occur in the physical system, the equations used to represent the physical system may accidentally contain ridges.

## 9-13 STEEPEST-ASCENT METHOD

As the name implies, this multivariable search method moves the state point in such a direction that the objective function changes at the greatest favorable rate. As shown in Sec. 8-7, the gradient vector is normal to the contour line or surface and therefore indicates the direction of maximum rate of change. In the function of two variables whose contour lines are shown in Fig. 9-11, the gradient vector $\nabla y$ at point $A$ is normal to the contour line at that point and indicates the direction in which $y$ increases at the greatest rate with respect to distance on the $x_1 x_2$ plane. The equation for $\nabla y$ is

$$\nabla y = \frac{\partial y}{\partial x_1} i_1 + \frac{\partial y}{\partial x_2} i_2 \tag{9-7}$$

where $i_1$ and $i_2$ are unit vectors in the $x_1$ and $x_2$ directions, respectively.

The essential steps in executing the steepest-ascent method are as follows:

1. Select a trial point.
2. Evaluate the gradient at the current point and the relationship of the changes of the $x$ variables.
3. Decide in which direction to move along the gradient.
4. Decide how far to move and then move that distance.

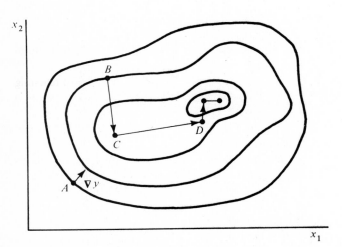

**Figure 9-11** Steepest-ascent method.

5. Test to determine whether the optimum has been achieved. If so, terminate, otherwise return to step 2.

The first three steps are standard, but there are many variations of steps 4 and 5. The individual steps will now be discussed.

**Step 1** The trial point should be chosen as near to the optimum as possible, but usually such insight is not available and the point is selected arbitrarily.

**Step 2** The partial derivatives can be extracted mathematically, or it may be more convenient to compute them numerically by resorting to the equation for the partial derivative,

$$\frac{\partial y}{\partial x_i} = \frac{y(x_1, \ldots, x_i + \Delta, \ldots, x_n) - y(x_1, \ldots, x_i, \ldots, x_n)}{\Delta} \qquad (9\text{-}8)$$

where $\Delta$ is a very small value. In order to move in the direction of the gradient, the relationship of the changes in the $x$'s, called $\Delta x$'s, is

$$\frac{\Delta x_1}{\partial y/\partial x_1} = \frac{\Delta x_2}{\partial y/\partial x_2} = \cdots = \frac{\Delta x_n}{\partial y/\partial x_n} \qquad (9\text{-}9)$$

**Step 3** It is understood that $\Delta x_i$ indicates the change in $x_i$, but a decision must be made whether to increase $x_i$ by $\Delta x_i$ or to decrease $x_i$. That decision is controlled by whether a maximization or minimization is being performed. If a maximization is in progress and $\partial y/\partial x_i$ is negative, $\Delta x_i$ must be negative in order for $y$ to increase as a result of the move.

**Step 4** Only two of the numerous methods for deciding how far to move in the direction of steepest ascent will be presented. The first method is to select arbitrarily a step size for one variable, say $\Delta x_1$, and compute $\Delta x_2$ to $\Delta x_n$ from Eq. (9-9). This method usually works well until one or more of the partial derivatives approaches zero.

The second method of deciding how far to move is to proceed until reaching an optimal value in the direction of the gradient. Then another gradient vector is calculated, and once again the position moves until an optimal value in that direction is achieved. From point $B$ in Fig. 9-11 the point moves to $C$ and thence to $D$. The process of determining the optimal value in the gradient direction can be converted into a single-variable search by using Eq. (9-9) to eliminate all $x$'s except one in the objective function. The Fibonacci method can then be used as an efficient single-variable search.

**Example 9-2** An insulated steel tank storing ammonia, as shown in Fig. 9-12, is equipped with a recondensation system which can control the pressure and thus the temperature of the ammonia.[3] Two basic decisions

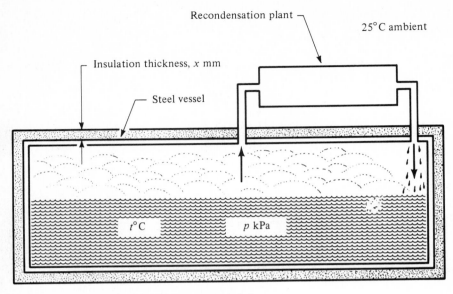

**Figure 9-12** Ammonia storage tank in Example 9-2.

to make in the design of the tank are the shell thickness and insulation thickness.

If the tank operates with a temperature near ambient, the pressure in the tank will be high and a heavy expensive vessel will be required. On the other hand, to maintain a low pressure in the tank requires more operation of the recondensation system because there will be more heat transferred from the environment unless the insulation is increased, which also adds cost.

Determine the optimum operating temperature and insulation thickness if the following costs and other data apply:

Vessel cost, $1000 + 2.2(p - 100)^{1.2}$ dollars for $p \geqslant 200$ kPa
Insulation cost for the 60 m$^2$ of heat-transfer area, $21x^{0.9}$ dollars
Recondensation cost, 2.5 cents per kilogram of ammonia
Lifetime hours of operation, 50,000 h
Ambient temperature, 25°C
Average latent heat of vaporization of ammonia, 1200 kJ/kg
Conductivity of insulation $k$, 0.04 W/(m · K)
Pressure-temperature relation for ammonia

$$\ln p = -\frac{2800}{t + 273} + 16.33$$

SOLUTION The total lifetime cost is the sum of three individual costs, the vessel, the insulation, and the lifetime cost of recondensation. All these costs will be expressed in terms of the operating temperature $t°C$ and the insulation thickness $x$ mm.

The insulation cost is

$$IC = 21x^{0.9}$$

The saturation pressure is a function of temperature

$$p = e^{-2800/(t+273)+16.33}$$

and so the vessel cost VC is

$$VC = 1000 + 2.2\,(e^{-2800/(t+273)+16.33} - 100)^{1.2}$$

Recondensation cost RC is

$$RC = (w \text{ kg/s})\,(0.025 \text{ \$/kg})\,(3600 \text{ s/h})\,(50{,}000 \text{ h}) \qquad (9\text{-}10)$$

where $w$ is the evaporation and recondensation rate in kilograms per second. But also

$$w = \frac{q \text{ kW}}{1200 \text{ kJ/kg}} \qquad (9\text{-}11)$$

where $q$ is the rate of heat transfer from the environment to the ammonia. Assuming that only the insulation provides any significant resistance to heat transfer, we have

$$q, \text{kW} = \frac{25 - t}{(x \text{ mm})/1000}\,[0.00004 \text{ kW/(m} \cdot \text{K)}]\,(60 \text{ m}^2) \qquad (9\text{-}12)$$

Combining Eqs. (9-10) to (9-12) results in the expression for the recondensation cost

$$RC = \frac{9000\,(25 - t)}{x}$$

The total cost is the sum of the individual costs,

$$\text{Total cost } C = IC + VC + RC$$

The search method chosen to perform this optimization will be the steepest descent. The position will be moved along the gradient direction until a minimum is reached; then a new gradient will be established. The partial derivatives of the total cost $C$ with respect to $x$ and $t$ are

$$\frac{\partial C}{\partial x} = (0.9)\,(21)x^{-0.1} - \frac{9000\,(25 - t)}{x^2}$$

$$\frac{\partial C}{\partial t} = (2.2)\,(1.2)\,(e^A - 100)^{0.2}e^A\,\frac{2800}{(t + 273)^2} - \frac{9000}{x}$$

**TABLE 9-1 Steepest-descent search in Example 9-2**

| Iteration | $x$, mm | $t$, °C | $C$ | $\partial C/\partial x$ | $\partial C/\partial t$ |
|-----------|---------|---------|-----|-------------------------|-------------------------|
| 0 | 100.00 | 5.00 | $7237.08 | −6.075 | 77.347 |
| 1 | 101.23 | −10.66 | 6675.58 | −19.409 | −1.455 |
| 2 | 142.83 | −7.54 | 6334.98 | −2.845 | 37.505 |
| 3 | 143.62 | −17.78 | 6152.85 | −7.165 | −0.634 |
| 4 | 168.92 | −15.54 | 6067.68 | −1.471 | 16.145 |
| 5 | 169.40 | −20.80 | 6026.09 | −3.051 | −0.294 |
| 6 | 182.87 | −19.50 | 6005.85 | −0.749 | 7.477 |
| . . . . . . . . . . . . . . . . . . . . . . . . . . . . . . . . . . . . . . . . . . . . . . . . . |
| 32 | 196.56 | −23.24 | 5986.07 | −0.090 | 0.189 |
| 33 | 196.62 | −23.36 | 5986.05 | −0.112 | −0.123 |
| 34 | 196.69 | −23.28 | 5986.05 | −0.086 | 0.111 |
| 35 | 196.77 | −23.38 | 5986.04 | −0.100 | −0.143 |
| 36 | 196.83 | −23.30 | 5986.04 | −0.076 | 0.088 |

where
$$A = -\frac{2800}{t + 273} + 16.33$$

For the first point, arbitrarily select $x$ = 100 mm and $t$ = 5°C. At this position $C$ = \$7237.08, $\partial C/\partial x$ = − 6.075, and $\partial C/\partial t$ = 77.347. The derivative with respect to $x$ is negative and with respect to $t$ is positive; therefore to decrease $C$ the value of $x$ must be increased and $t$ must be decreased. Furthermore, to move along the direction of the gradient, the changes in $x$ and $t$, designated $\Delta x$ and $\Delta t$, should bear the relation

$$\frac{\Delta x}{-6.075} = \frac{\Delta t}{77.347}$$

The minimum value of $C$ achieved by moving in the direction of this gradient is \$6675.58, where $x$ = 101.23 and $t$ = − 10.66. New partial derivatives are computed at this point and the position changed according to the new gradient. Table 9-1 presents a summary of the calculations.

The minimum cost is \$5986.04 when the insulation thickness is 196.8 mm and the operating temperature is − 23.3°C. Both partial derivatives are nearly zero at this point. The steepest-descent calculation required an appreciable number of steps before finally homing in on the optimum. The reason is that the route passes through a curved valley, and the minimum point along the gradient moved from one side of the valley to the other, as indicated by the alternation in sign of $\partial C/\partial t$.

## 9-14 SCALES OF THE INDEPENDENT VARIABLES

The name steepest ascent implies the best possible direction in which to move. The meaning of this statement is that for a given distance $\sqrt{\Delta x_1^2 + \Delta x_2^2 + \cdots}$ the

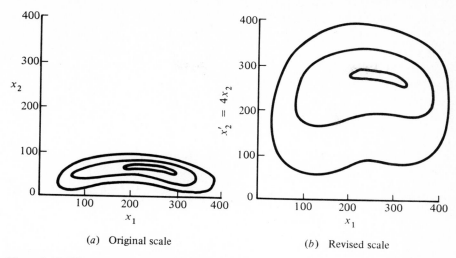

(a) Original scale    (b) Revised scale

**Figure 9-13** Effect of scale of independent variables.

objective function will experience a maximum change. As Fig. 9-13a shows, however, it may be desirable for there to be large changes in $x_1$ compared with those of $x_2$. Wilde[2] extends the conclusions of Buehler, Shah, and Kempthorne[4] to recommend that the scales be chosen so that the contours are as spherical as possible in order to accelerate the convergence. In Fig. 9-13b, for example, the original equation would be revised with a new variable $x_2'$ replacing $x_2$ so that $x_2' = 4x_2$ and the contours would thus cover the same range as $x_1$.

## 9-15 CONSTRAINED OPTIMIZATION

Constrained optimizations are probably the most frequent and most important ones encountered in the design of thermal systems. Numerous methods and variations of search methods applicable to constrained optimizations are presented in the literature, most of them prompted by attempts to accelerate the search. In some cases new methods were developed to prevent the search from failing on a certain function or problem. It is this multitude of techniques that blunts the satisfaction of using search methods, because there is not just one fundamental concept, as there is with Lagrange multipliers.

Two methods of constrained searches will be presented in the next several sections, (1) conversion to unconstrained by use of penalty functions and (2) searching along the constraint. Only equality constraints (and not inequality constraints) will be considered, although the user of a search could adapt the equality-constraint technique by making an inequality constraint active or inactive depending upon whether the constraint is violated or not.

## 9-16 PENALTY FUNCTIONS

If a function is to be maximized

$$y = y(x_1, x_2, \ldots, x_n) \longrightarrow \text{maximum}$$

subject to the constraints

$$\phi_1(x_1, x_2, \ldots, x_n) = 0$$
$$\cdots\cdots\cdots\cdots\cdots\cdots$$
$$\phi_m(x_1, x_2, \ldots, x_n) = 0$$

a new unconstrained function can be constructed

$$Y = y - P_1(\phi_1)^2 - P_2(\phi_2)^2 - \cdots - P_m(\phi_m)^2$$

If the function is to be minimized, the $P_i(\phi_i)^2$ terms would be added to the original objective function to construct the new unconstrained function. The underlying principle of the technique is valid, but care must be exercised in the execution to maintain proper relative influence of the function being optimized to that of the constraints. The choice of the $P$ terms provides the relative weighting of the two influences, and if $P$ is too high, the search will satisfy the constraint but move very slowly in optimizing the function. If $P$ is too small, the search may terminate without satisfying the constraints adequately. One suggestion is to start with small values of the $P$'s and gradually increase the values as the magnitudes of the $\phi$'s become small.

## 9-17 OPTIMIZATION BY SEARCHING ALONG A CONSTRAINT

The next search technique for constrained optimization to be explained is the *search along the constraint* or *hemstitching* method.[5,6] The technique consists of starting at a trial point and first driving directly toward the constraint(s). Once on the constraint(s), the process is one of optimizing along the constraint. For nonlinear constraints a tangential move starting on a constraint moves slightly off the constraint, so after each tangential move it is necessary to drive back onto the constraint.

This search method is one of many that are available but is effective in most problems and offers the further satisfaction of building logically on the principles from the calculus method.

## 9-18 DRIVING TOWARD A SINGLE CONSTRAINT

The preliminary operation in the search-along-the-constraint method is to locate a point on the constraint. Perhaps the constraint will be simple enough to be solved explicitly for one variable, in which case arbitrary values can be chosen for all but one variable and the constraint solved for that variable. We are particularly interested, however, in cases where the constraint is complex. In these cases a form of the steepest-descent search can be used. If the problem consists of two variables of optimization, $x_1$ and $x_2$, and the constraint is $\phi(x_1, x_2) = 0$, suppose that the trial values of $x_1$ and $x_2$ are $a_1$ and $a_2$, respectively. Probably $a_1$ and $a_2$ do not satisfy the constraint, but instead

$$\phi(a_1, a_2) = b \qquad (9\text{-}13)$$

An effective direction in which to move is indicated by the gradient vector

$$\nabla\phi = \frac{\partial\phi}{\partial x_1} \mathbf{i}_1 + \frac{\partial\phi}{\partial x_2} \mathbf{i}_2 \qquad (9\text{-}14)$$

which means that $x_1$ and $x_2$ are to be incremented according to the relation

$$\frac{\Delta x_1}{\partial\phi/\partial x_1} = \frac{\Delta x_2}{\partial\phi/\partial x_2} \qquad (9\text{-}15)$$

Furthermore an estimate of the extent of the move is available from

$$\Delta\phi \approx \frac{\partial\phi}{\partial x_1} \Delta x_1 + \frac{\partial\phi}{\partial x_2} \Delta x_2 = -b \qquad (9\text{-}16)$$

The terms in Eq. (9-16) are equated to $-b$ in order to bring the value of $\phi$ back to zero. The combination of Eqs. (9-15) and (9-16) permits solution for one of the $\Delta x$'s, for example $\Delta x_1$,

$$\Delta x_1 \left[ \frac{\partial\phi}{\partial x_1} + \frac{(\partial\phi/\partial x_2)^2}{\partial\phi/\partial x_1} \right] = -b \qquad (9\text{-}17)$$

Equation (9-17) is solved for $\Delta x_1$, which is then substituted into Eq. (9-15) to compute $\Delta x_2$.

**Example 9-3** A constraint in an optimization problem is $x_1^2 x_2 = 8$. Arbitrarily choosing trial values $x_1 = 2$ and $x_2 = 1$, use the gradient vector to drive toward the constraint.

**Solution** The expression for $\phi$ is

$$\phi = x_1^2 x_2 - 8$$

and when $x_1 = 2$ and $x_2 = 1$, $\phi = -4$. The partial derivatives are

$$\frac{\partial\phi}{\partial x_1} = 2x_1 x_2 = 4 \quad \text{and} \quad \frac{\partial\phi}{\partial x_2} = x_1^2 = 4$$

Computing $\Delta x_1$ from Eq. (9-17) yields

$$\Delta x_1 = \frac{-(-4)}{4 + 4^2/4} = 0.5$$

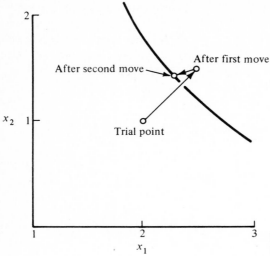

**Figure 9-14** Driving toward a constraint in Example 9-3.

and from Eq. (9-15) $\Delta x_2 = 0.5$. The new point is $x_1 = 2.5$ and $x_2 = 1.5$, as shown in Fig. 9-14.

Figure 9-14 shows the direction of the move to have been quite favorable. The direction was not precisely toward the nearest point on the constraint, however. The direction was normal to $\phi$ at the trial point. The calculated extent of the move was not quite correct, and the new point is not exactly on the constraint. The reason for the imprecision is that the partial derivatives are applicable only to the trial point $(2, 1)$. The value of $\phi$ is now 1.375. If the process is repeated from the current point $(2.5, 1.5)$, the next point is $(2.392, 1.410)$ and $\phi = 0.068$.

The technique extends logically to more than two variables. If the current position is $x_1 = a_1$, $x_2 = a_2$, and $x_3 = a_3$, at which point the value of $\phi$ is $b$ rather than the sought-for value of zero, the change in variable $\Delta x_1$ is

$$\Delta x_1 = \frac{-b}{\partial \phi / \partial x_1 + (\partial \phi / \partial x_2)^2 / (\partial \phi / \partial x_1) + (\partial \phi / \partial x_3)^2 / (\partial \phi / \partial x_1)} \quad (9\text{-}18)$$

Then $\Delta x_2$ and $\Delta x_3$ are found from the equation of the form of Eq. (9-15).

## 9-19 DRIVING TOWARD ONE CONSTRAINT ALONG ANOTHER CONSTRAINT

The previous section explained how to move freely in the optimization space in a favorable direction toward one constraint. When there are two constraints, the progression toward satisfying both of them can be as follows: (1) from the trial point drive freely toward the first constraint, (2) when on, or sufficiently close to, the first constraint, drive *along* the first constraint toward the second constraint, and (3) drive along the second constraint back to the first one, iterating back and forth as necessary.

For the case of two constraints with three variables step 1 is executed as described in Sec. 9-18. Step 2 consists of driving $\phi_2$ to zero while moving tangent to constraint 1. The change in $\phi_2$, called $\Delta\phi_2$, is given by

$$\Delta\phi_2 = \frac{\partial\phi_2}{\partial x_1}\Delta x_1 + \frac{\partial\phi_2}{\partial x_2}\Delta x_2 + \frac{\partial\phi_2}{\partial x_3}\Delta x_3 \qquad (9\text{-}19)$$

The $\Delta x$'s are no longer free to move independently, but they are restricted by constraint 1, which requires that $\Delta\phi_1$ be zero in any move,

$$\Delta\phi_1 = \frac{\partial\phi_1}{\partial x_1}\Delta x_1 + \frac{\partial\phi_1}{\partial x_2}\Delta x_2 + \frac{\partial\phi_1}{\partial x_3}\Delta x_3 = 0 \qquad (9\text{-}20)$$

A further demand of the move toward the $\phi_2$ constraint is that it be accomplished in the most favorable direction, which specifies that $\Delta\phi_2$ change by a maximum amount for a shift of a given distance in the $x_1 x_2 x_3$ space. This requirement can be expressed as a constraint

$$\Delta x_1^2 + \Delta x_2^2 + \Delta x_3^2 = \text{const} \qquad (9\text{-}21)$$

Equations (9-20) and (9-21) are the constraints on the objective function, Eq. (9-19), for which optimization the Lagrange multiplier equations are

$\Delta x_1$:
$$\frac{\partial\phi_2}{\partial x_1} - \lambda_1 \frac{\partial\phi_1}{\partial x_1} - \lambda_2(2)\Delta x_1 = 0 \qquad (9\text{-}22)$$

$\Delta x_2$:
$$\frac{\partial\phi_2}{\partial x_2} - \lambda_1 \frac{\partial\phi_1}{\partial x_2} - \lambda_2(2)\Delta x_2 = 0 \qquad (9\text{-}23)$$

$\Delta x_3$:
$$\frac{\partial\phi_2}{\partial x_3} - \lambda_1 \frac{\partial\phi_1}{\partial x_3} - \lambda_2(2)\Delta x_3 = 0 \qquad (9\text{-}24)$$

Multiply Eq. (9-22) by $\partial\phi_1/\partial x_1$, Eq. (9-23) by $\partial\phi_1/\partial x_2$, and Eq. (9-24) by $\partial\phi_1/\partial x_3$; then total the first terms, second terms, and third terms of the three equations,

$$\left(\frac{\partial\phi_1}{\partial x_1}\frac{\partial\phi_2}{\partial x_1} + \frac{\partial\phi_1}{\partial x_2}\frac{\partial\phi_2}{\partial x_2} + \frac{\partial\phi_1}{\partial x_3}\frac{\partial\phi_2}{\partial x_3}\right) - \lambda_1\left[\left(\frac{\partial\phi_1}{\partial x_1}\right)^2 + \left(\frac{\partial\phi_1}{\partial x_2}\right)^2 + \left(\frac{\partial\phi_1}{\partial x_3}\right)^2\right]$$

$$- 2\lambda_2\left(\frac{\partial\phi_1}{\partial x_1}\Delta x_1 + \frac{\partial\phi_1}{\partial x_2}\Delta x_2 + \frac{\partial\phi_1}{\partial x_3}\Delta x_3\right) = 0 \quad (9\text{-}25)$$

A comparison of the last term in Eq. (9-25) with Eq. (9-20) shows the expression within the parentheses to be zero. For convenience, call

$$A = \frac{\partial\phi_1}{\partial x_1}\frac{\partial\phi_2}{\partial x_1} + \frac{\partial\phi_1}{\partial x_2}\frac{\partial\phi_2}{\partial x_2} + \frac{\partial\phi_1}{\partial x_3}\frac{\partial\phi_2}{\partial x_3}$$

and
$$B = \left(\frac{\partial \phi_1}{\partial x_1}\right)^2 + \left(\frac{\partial \phi_1}{\partial x_2}\right)^2 + \left(\frac{\partial \phi_1}{\partial x_3}\right)^2$$

Equation (9-25) then becomes

$$\lambda_1 = \frac{A}{B} \tag{9-26}$$

Substituting Eq. (9-26) into Eqs. (9-22) to (9-24) and solving for $1/2\lambda_2$ in each of the equations results in

$$\frac{1}{2\lambda_2} = \frac{\Delta x_1}{\partial \phi_2/\partial x_1 - (A/B)\,\partial \phi_1/\partial x_1} = \frac{\Delta x_2}{\partial \phi_2/\partial x_2 - (A/B)\,\partial \phi_1/\partial x_2}$$

$$= \frac{\Delta x_3}{\partial \phi_2/\partial x_3 - (A/B)\,\partial \phi_1/\partial x_3} \tag{9-27}$$

If the value of $\phi_2$ after reaching the first constraint is indicated by $b_2$, the change in one variable, $x_1$ for example, would be found by combining Eqs. (9-19) and (9-27),

$$\Delta x_1 = \frac{-b_2}{\dfrac{\partial \phi_2}{\partial x_1} + \dfrac{\partial \phi_2}{\partial x_2}\dfrac{\dfrac{\partial \phi_2}{\partial x_2} - \dfrac{A}{B}\dfrac{\partial \phi_1}{\partial x_2}}{\dfrac{\partial \phi_2}{\partial x_1} - \dfrac{A}{B}\dfrac{\partial \phi_1}{\partial x_1}} + \dfrac{\partial \phi_2}{\partial x_3}\dfrac{\dfrac{\partial \phi_2}{\partial x_3} - \dfrac{A}{B}\dfrac{\partial \phi_1}{\partial x_3}}{\dfrac{\partial \phi_2}{\partial x_1} - \dfrac{A}{B}\dfrac{\partial \phi_1}{\partial x_1}}} \tag{9-28}$$

The change in variables $x_2$ and $x_3$ is computed from Eq. (9-27).

In the process of executing step 2 the point may have moved off of the first constraint, and so it is now necessary in step 3 to drive back to constraint 1 along constraint 2. This operation uses Eq. (9-28) with the subscripts 1 and 2 interchanged in $b$ and in the $\phi$'s.

**Example 9-4** Two constraints exist in a three-variable problem,

$$x_1 x_2 x_3 - 8 = 0$$

$$x_1 x_2 - 2x_2 x_3 + 3x_1 x_3 - 12 = 0$$

Starting with the point (3, 3, 3), drive toward a point that satisfies both constraints.

SOLUTION   Step 1 drives in the gradient direction toward the first constraint, $\phi_1$, using Eq. (9-18),

$$\Delta x_1 = \frac{-19}{(3)(3) + 9^2/9 + 9^2/9} = -0.7037$$

$$\Delta x_2 = (-0.7037)\frac{\partial \phi_1/\partial x_2}{\partial \phi_1/\partial x_1} = -0.7037$$

**Table 9-2  Driving toward constraints in Example 9-4**

| | After moving | | | | | |
|---|---|---|---|---|---|---|
| Along | Toward | $x_1$ | $x_2$ | $x_3$ | $\phi_1$ | $\phi_2$ |
| | | 2.021 | 2.021 | 2.021 | 0.2547 | −3.8311 |
| Constraint 1 | Constraint 2 | 2.420 | 1.672 | 1.971 | −0.0251 | −0.2339 |
| Constraint 2 | Constraint 1 | 2.420 | 1.676 | 1.973 | 0.0000 | −0.2340 |

and
$$\Delta x_3 = -0.7037$$

The new values are $x_1 = x_2 = x_3 = 2.296$. The new $\phi_1$ is 4.10. Another step brings the following values of the $x$'s:

$$x_1 = 2.021 \qquad x_2 = 2.021 \qquad x_3 = 2.021 \qquad \phi_1 = 0.255$$

Assume that this position is adequately close to constraint 1 for the present.

The information needed for step 2 is $b_2 = -3.831$, $\partial\phi_2/\partial x_1 = 8.084$, $\partial\phi_2/\partial x_2 = -2.021$, $\partial\phi_2/\partial x_3 = 2.021$, $\partial\phi_1/\partial x_1 = 4.084$, $\partial\phi_1/\partial x_2 = 4.084$, $\partial\phi_1/\partial x_3 = 4.084$:

$$A = (4.084)(8.084) + (4.084)(-2.021) + (4.084)(2.021) = 33.02$$
$$B = (4.084)^2 + (4.084)^2 + (4.084)^2 = 50.05$$

From Eq. (9-28)

$$\Delta x_1 = \frac{-(-3.831)}{8.084 + (-2.021)(-4.716/5.389) + (2.021)(-0.6737/5.389)}$$
$$= 0.3991$$

and from Eq. (9-27),

$$\Delta x_2 = \frac{(0.3991)(-4.716)}{5.389} = -0.3492$$

$$\Delta x_3 = \frac{(0.3991)(-0.6737)}{5.389} = -0.0499$$

The summary of the iterations as they draw the position toward the constraints is presented in Table 9-2.

## 9-20  OPTIMIZATION ALONG A CONSTRAINT

Sections 9-18 and 9-19 prepare for the final operation of the search method of optimization along a constraint. The complete process consists of starting at the trial point and then first driving to the constraint(s). Once on the constraint(s), the operations alternate between a short move along the constraint(s) in a direction that improves the objective function and then a return to the constraint. If any of the constraints are nonlinear, a move that starts tangent to the constraints will carry the position off the constraint, so it is necessary to return to the constraint in the most direct manner.

The process of moving in the direction of steepest ascent (descent) of the objective function is similar to moving on a constraint toward another constraint, as explained in Sec. 9-19. For example, if there are three variables of optimization and one constraint, the equations analogous to Eqs. (9-19), (9-20), and (9-27) are

$$\Delta y = \frac{\partial y}{\partial x_1} \Delta x_1 + \frac{\partial y}{\partial x_2} \Delta x_2 + \frac{\partial y}{\partial x_3} \Delta x_3 \tag{9-29}$$

$$\Delta \phi = \frac{\partial \phi}{\partial x_1} \Delta x_1 + \frac{\partial \phi}{\partial x_2} \Delta x_2 + \frac{\partial \phi}{\partial x_3} \Delta x_3 = 0 \tag{9-30}$$

$$\frac{\Delta x_1}{\partial y/\partial x_1 - (A/B) \partial \phi/\partial x_1} = \frac{\Delta x_2}{\partial y/\partial x_2 - (A/B) \partial \phi/\partial x_2} = \frac{\Delta x_3}{\partial y/\partial x_3 - (A/B) \partial \phi/\partial x_3} \tag{9-31}$$

where $A = \dfrac{\partial y}{\partial x_1} \dfrac{\partial \phi}{\partial x_1} + \dfrac{\partial y}{\partial x_2} \dfrac{\partial \phi}{\partial x_2} + \dfrac{\partial y}{\partial x_3} \dfrac{\partial \phi}{\partial x_3}$

$$B = \left(\frac{\partial \phi}{\partial x_1}\right)^2 + \left(\frac{\partial \phi}{\partial x_2}\right)^2 + \left(\frac{\partial \phi}{\partial x_3}\right)^2$$

One difference that arises between driving toward another constraint, as in Sec. 9-19, and in improving the objective function is that the magnitude of the move must be chosen by the searcher. Noticeable progress must be made without driving too far away from the constraint. Furthermore, it is not possible to optimize in the direction indicated by Eq. (9-31), because the optimum in that direction may be so far off the constraint that the values are meaningless.

**Example 9-5** The objective function associated with the constraint in Example 9-3, $x_1^2 x_2 - 8 = 0$, is

$$y = 3x_1^2 + x_2^2$$

Minimize this function by searching along the constraint, choosing $\Delta x_2$ increments of 0.1, starting at the position achieved in Example 9-3: $x_1 = 2.392$ and $x_2 = 1.410$.

SOLUTION A change in $y$, designated $\Delta y$, can be expressed by

$$\Delta y = \frac{\partial y}{\partial x_1} \Delta x_1 + \frac{\partial y}{\partial x_2} \Delta x_2 \tag{9-32}$$

but the changes in position $\Delta x_1$ and $\Delta x_2$ are to be related so that the motion is tangent to the constraint; thus

$$\Delta x_1 = - \frac{\partial \phi / \partial x_2}{\partial \phi / \partial x_1} \Delta x_2 \qquad (9\text{-}33)$$

Substituting Eq. (9-33) into Eq. (9-32) gives

$$\Delta y = \left( \frac{\partial y}{\partial x_2} - \frac{\partial y}{\partial x_1} \frac{\partial \phi / \partial x_2}{\partial \phi / \partial x_1} \right) \Delta x_2 \qquad (9\text{-}34)$$

The magnitude of $\Delta x_2$ is specified as 0.1, but the sign must be such that $y$ decreases for the minimization. Designating $G$ as the term in the parentheses in Eq. (9-34), we have

$$\Delta y = G \, \Delta x_2$$

so for minimization

$$\Delta x_2 = - \frac{G}{\text{absolute magnitude of } G} (0.1)$$

After computation of $\Delta x_2$, $\Delta x_1$ is available from Eq. (9-33). Following the shift tangent to the constraint, the position is brought back onto the constraint using Eqs. (9-16) and (9-17).

Table 9-3 shows the progression of the search, and Fig. 9-15 illustrates the first several tangent moves and the returns to the constraint.

As the position moved from (1.901, 2.215) to (1.859, 2.315), the optimum was passed and the search then reversed its direction. If a more precise location

**Table 9-3  Search along constraint in Example 9-5**

| Cycle | After move | $x_1$ | $x_2$ | $y$ | $\phi$ | $G$ |
|-------|------------|-------|-------|-----|--------|-----|
|       | (Start)    | 2.392 | 1.410 | 19.153 | 0.0675 | −9.35 |
| 1     | Tangent    | 2.307 | 1.510 | 18.249 | 0.0378 |       |
|       | Return     | 2.304 | 1.507 | 18.194 | 0.0001 | −7.55 |
| 2     | Tangent    | 2.227 | 1.607 | 17.467 | −0.0258 |      |
|       | Return     | 2.230 | 1.609 | 17.505 | 0.0000 | −6.05 |
| 3     | Tangent    | 2.160 | 1.709 | 16.924 | −0.0227 |      |
|       | Return     | 2.163 | 1.710 | 16.957 | 0.0000 | −4.78 |
| . . . | . . . . . . . . | . . . | . . . | . . . | . . . | . . . |
| 8     | Tangent    | 1.899 | 2.214 | 15.724 | −0.0132 |      |
|       | Return     | 1.901 | 2.215 | 15.742 | 0.0000 | −0.46 |
| 9     | Tangent    | 1.858 | 2.315 | 15.711 | −0.0121 |      |
|       | Return     | 1.859 | 2.315 | 15.726 | 0.0000 | +0.15 |
| 10    | Tangent    | 1.899 | 2.215 | 15.726 | −0.0114 |      |
|       | Return     | 1.900 | 2.216 | 15.741 | 0.0000 | −0.46 |
| 11    | Tangent    | 1.857 | 2.316 | 15.711 | −0.0120 |      |

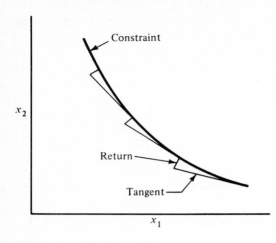

**Figure 9-15** Search along the constraint (hemstitching) in Example 9-5.

of the optimum were desired, the step size of $\Delta x_2$ could have been reduced at this point.

Associated with the movement to and even slightly past the optimum in Example 9-5 was the reversal of $G$ from negative values to positive values. In other words $G$ reached the value of zero at the optimum; thus

$$\frac{\partial y}{\partial x_2} - \frac{\partial y}{\partial x_1} \frac{\partial \phi / \partial x_2}{\partial \phi / \partial x_1} = 0 \qquad (9\text{-}35)$$

This relationship is predictable because the Lagrange multiplier equations would require for the optimum that

$$\frac{\partial y}{\partial x_1} - \lambda \frac{\partial \phi}{\partial x_1} = 0 \quad \text{and} \quad \frac{\partial y}{\partial x_2} - \lambda \frac{\partial \phi}{\partial x_2} = 0$$

or

$$\lambda = \frac{\partial y / \partial x_1}{\partial \phi / \partial x_1} = \frac{\partial y / \partial x_2}{\partial \phi / \partial x_2}$$

which is equivalent to Eq. (9-35).

The process of searching along the constraint followed in this search technique is also called hemstitching, which comes from the appearance of the search progression shown in Fig. 9-15. The sequence is to move alternately tangent to the constraint in a direction that changes the objective function favorably and then return as directly as possible to the constraint.

## 9-21 SUMMARY

This chapter explored single-variable searches as well as both unconstrained and constrained multivariable searches. The types of problems for which search methods are most likely to be called into service are the difficult ones, which are probably the multivariable constrained problems. Of the many methods available for multivariable constrained optimizations, the search along the constraint(s) was chosen because of its wide applicability (although it is not necessarily the most efficient) and because it follows logically from calculus methods.

The actual execution of the calculations in complicated problems would probably be carried out on a computer, and the availability of the interactive mode is particularly convenient. The possibility the searcher has of changing such quantities as the step size or starting with a new trial point facilitates convergence to the optimum in a rapid manner.

This chapter emphasized equality constraints, but the extension to inequalities can be achieved by testing the inequality constraint at each new position. If the inequality is not violated, the constraint is ignored; if it is violated, it can then be included as an equality constraint.

## PROBLEMS

**9-1** The function

$$y = \frac{(\ln x) \sin (x^2/25)}{x}$$

where $x^2/25$ is in radians, is unimodal in the range $1.5 \leqslant x \leqslant 10$.

(a) If a Fibonacci search is employed to determine the maximum, how many points will be needed for the final interval of uncertainty of $x$ to be 0.3 or less?

(b) Using the number of points determined in part (a), conduct the Fibonacci search and determine the interval in which the maximum occurs.

**Ans.:** (b) 5.75 to 6.00.

**9-2** One of the strategies in some search methods is to use a coarse subdivision first to determine the approximate region of the optimum and then a fine subdivision for a second search. For a single-variable search 16 points total are to be applied. Compare the ratio of the initial to final interval of uncertainty if (a) all 16 points are used in one Fibonacci search and (b) 8 points in a Fibonacci search are used to determine an interval of uncertainty of reduced size on which another 8-point Fibonacci search is applied.

**9-3** An economic analysis of a proposed facility is being conducted to select an operating life such that the maximum uniform annual income is achieved. A short life results in high annual amortization costs, but the maintenance costs become excessive for a long life. The annual income after deducting all operating expenses, except maintenance costs, is $180,000. The first cost of the facility is $500,000 borrowed at 10 percent interest compounded annually.

The maintenance costs are zero at the end of the first year, $10,000 at the end of the second, $20,000 at the end of the third, etc. To express these maintenance charges on an annual basis the gradient present-worth factor of Sec. 3-8 can be multiplied by the capital-recovery factor, which for the 10 percent interest is presented in Table 9-4.

Use a Fibonacci search for integer years between 0 and 21 to find the life of the facility which results in the maximum annual profit. Omit the last calculation of the Fibonacci process since we are interested only in integer-year results.

**Ans.:** 12 years, $62,760 annual income.

**Table 9-4 Factors for conversion of gradient series to an annual cost**

| Year | Factor | Year | Factor | Year | Factor |
|------|--------|------|--------|------|--------|
| 1 | 0.000 | 8 | 3.008 | 15 | 5.275 |
| 2 | 0.476 | 9 | 3.376 | 16 | 5.552 |
| 3 | 0.937 | 10 | 3.730 | 17 | 5.801 |
| 4 | 1.379 | 11 | 4.060 | 18 | 6.058 |
| 5 | 1.810 | 12 | 4.384 | 19 | 6.295 |
| 6 | 2.224 | 13 | 4.696 | 20 | 6.500 |
| 7 | 2.622 | 14 | 5.002 | 21 | 6.703 |

**9-4** The exhaust-gas temperature leaving a continuously operating furnace is $260°C$, and a proposal is being considered to install a heat exchanger in the exhaust-gas stream to generate low-pressure steam at $105°C$. The question to be investigated is whether it is economical to install such a heat exchanger, and, if so, to find its optimum size. The following data apply:

Flow of exhaust gas, 7.5 kg/s.
Specific heat of exhaust gas, 1.05 kJ/(kg · K).
Value of the heat in the form of steam, $1.50 per gigajoule.
$U$ value of heat exchanger based on gas-side area, 23 W/(m$^2$ · K).
Cost of the heat exchanger including installation based on gas-side area, $90 per square meter.
Interest rate, 8 percent.
Life of the installation, 5 years.
Saturated liquid water enters heat exchanger at $105°C$ and leaves as saturated vapor.

 (a) Develop the equation for the savings as a function of the area, expressed as a uniform annual amount.
 (b) What is the maximum permitted area if the exit-gas temperature is to be above $120°C$ in order to prevent condensation of water vapor from the exhaust gas?
 (c) Use a seven-point Fibonacci search and set up a table to simplify calculation of the optimum heat-transfer area.
 **Ans.:** Optimum area between 686 and 724 m$^2$.

**9-5** Perform a univariate search to find the minimum value of the function

$$y = x_1 + \frac{16}{x_1 x_2} + \frac{x_2}{2}$$

using only integer values of $x_1$ and $x_2$ and starting with $x_2 = 3$.
 **Ans.:** $y^* = 6$.

**9-6** The minimum value of the function

$$y = \frac{72 x_1}{x_2} + \frac{360}{x_1 x_3} + x_1 x_2 + 2 x_3$$

is to be sought using the method of steepest descent. If the starting point is $x_1 = 5, x_2 = 6$, $x_3 = 8$ ($y = 115$ at this point) and $x_1$ is to be changed by 1.0, what is the location of the next point?
 **Ans.:** New $y = 98.1$.

**9-7** The function

$$y = 2 x_1 x_2 + \frac{3}{x_1 x_2 x_3} + x_1 x_3^2 + 3 x_2$$

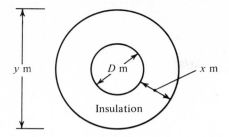

**Figure 9-16** Insulated pipe in Prob. 9-8.

is to be minimized by the steepest-descent search method. From the starting point the direction of steepest descent will be determined, and the search is to move in that direction until a minimum is reached, whereupon a new location is to be ascertained. If the arbitrarily chosen first point is (1, 1, 1) what is the second point?

    **Ans.:** $y$ at second point = 8.548.

**9-8** A pipe carrying high-temperature water is to be insulated and then mounted in a restricted space, as shown in Fig. 9-16. The choices of the pipe diameter $D$ m and the insulation thickness $x$ m are to be such that the OD of the insulation $y$ is a minimum, but the total annual operating cost of the installation is limited to \$40,000. This annual operating cost has two components, the water pumping cost and the cost of the heat loss:

$$\text{Pumping cost} = \frac{8}{D^5} \text{ dollars}$$

$$\text{Heat cost} = \frac{1500}{x} \text{ dollars}$$

    (*a*)  Write the objective function and the constraint.

    (*b*)  The values of $D = 0.2$ and $x = 0.1$ satisfy the constraint. Starting at this point move in a favorable direction tangent to the constraint by an increment of $\Delta D = 0.005$; then return in the most direct manner to the constraint. What are the new values of $x$ and $D$ following this hemstitch move?

    **Ans.:** $y = 0.3664$ m, $\phi = 19.86$.

**9-9** A refrigeration plant chills water and delivers it to a heat exchanger some distance away, as shown in Fig. 9-17. The supply water temperature is $t_1$°C, and the return water temperature is $t_2$. The flow rate is $w$ kg/s, and the pipe diameter is $D$ m. The cooling duty at the heat exchanger is 1200 kW, and the arithmetic mean temperature, $(t_1 + t_2)/2$, must be 12°C in order to transfer the 1200 kW from the air. The pump develops a pressure rise of 100,000 Pa that may be assumed independent of flow rate. The equation for pressure drop in a circular pipe is

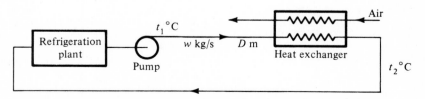

**Figure 9-17** Refrigeration plant, heat exchanger, and interconnecting piping.

$$\Delta p, \text{Pa} = f \frac{L}{D} \frac{V^2}{2} \rho$$

where $f = 0.02$
 $L = 2100$ m
 $\rho = 1000$ kg/m$^3$

The specific heat of water is 4.19 kJ/(kg · K).
 The three major costs associated with the choice of $t_1$, $w$, and $D$ are

Cost of the pipe, $150,000D$ dollars
Present worth of lifetime pumping costs, $500w$ dollars
Present worth of lifetime chilling costs, $60,000 - 4000t_1$ dollars

(a) Set up the equation for the cost as the objective function, the pressure drop as constraint 1, and the heat-transfer requirements as constraint 2.
 (b) Starting with a trial point of $w = 30$ kg/s, $t_1 = 5°$C, and $D = 0.2$ m, what is the resulting point after driving in the gradient direction toward constraint 1 and then tangent to constraint 1 in a direct manner toward constraint 2?
 (c) The point $w = 26$ kg/s, $D = 0.187$ m, and $t_1 = 6.5°$C essentially satisfies both constraints. What is the new point after making a move of $w = 1$ kg/s tangent to both constraints in a direction that reduces the cost?
 **Ans.:** (a) $(2.938 \times 10^6)D^5 - w^2 = 0$ and $w(24 - 2t_1) - 286.4 = 0$. (b) $D = 0.197$, $w = 29.507$, $t_1 = 7.11$. (c) $D = 0.184$, $w = 25$, $t_1 = 6.3$.

## REFERENCES

1. J. Kiefer, "Sequential Minimax Search for a Maximum," *Proc. Am. Math. Soc.*, vol. 4, p. 502, 1953.
2. D. J. Wilde, *Optimum Seeking Methods*, Prentice-Hall, Englewood Cliffs, N.J., 1964.
3. N. McCloskey, "Storage Facilities Associated with an Ammonia Pipeline," *ASME Pap.* 69-Pet-21, 1969.
4. R. J. Buehler, B. V. Shah, and C. Kempthorne: *Some Properties of Steepest Ascent and Related Procedures for Finding Optimum Conditions*, Iowa State University Statistical Laboratory, pp. 8-10, April 1961.
5. G. S. G. Beveridge and R. S. Schechter: *Optimization: Theory and Practice*, McGraw-Hill, New York, 1970.
6. S. M. Roberts and H. I. Lyvers, "The Gradient Method in Process Control," *Ind. Eng. Chem.*, vol. 53, pp. 877–882, 1961.

# TEN

## DYNAMIC PROGRAMMING

## 10-1 UNIQUENESS OF DYNAMIC-PROGRAMMING PROBLEMS

Dynamic programming is a method of optimization that is applicable either to staged processes or to continuous functions that can be approximated by staged processes. The word "dynamic" has no connection with the frequent use of the word in engineering terminology, where dynamic implies changes with respect to time.

As a method of optimization, dynamic programming is not usually interchangeable with such other forms of optimization as Lagrange multipliers and linear and nonlinear programming. Instead, it is related to the calculus of variations, whose result is an optimal *function* rather than an optimal *state point*. An optimization problem that can be subjected to dynamic programming or the calculus of variations is usually different from those suitable for treatment by Lagrange multipliers and linear and nonlinear programming. The calculus of variations is used, for example, to determine the trajectory (thus, a function in spatial coordinates) that results in minimum fuel cost of a spacecraft. Dynamic programming can attack this same problem by dividing the total path into a number of segments and considering the continuous function as a series of steps or stages. In such an application, the finite-step approach of dynamic programming is an approximation of the calculus-of-variations method. In many engineering situations, on the other hand, the problem consists of analysis of discrete stages, such as a series of compressors, heat exchangers, or reactors. These cases fit dynamic programming exactly, and here the calculus of variations would only approximate the result.

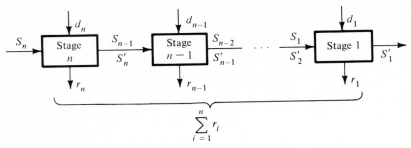

**Figure 10-1** Pictorial representation of problem that can be solved by dynamic programming; $S$ = state of the input to a stage, $S'$ = state of the output from a stage, $d$ = decision variable, and $r$ = return from a stage.

## 10-2 SYMBOLIC DESCRIPTION OF DYNAMIC PROGRAMMING

Figure 10-1 shows a symbolic description of the dynamic-programming problem. The decision variables are to be chosen so that for a specified input to stage $n$ and a specified output from stage 1 the summation of returns $\sum_{i=1}^{n} r_i$ is optimum (either maximum or minimum, depending upon the problem).

The description of the calculus of variations in Sec. 10-1 suggests that a function, for example $y(x)$, is sought. In Fig. 10-1 that goal is to find the optimal state variables $S$ for the various stages, where the stage corresponds to the $x$ variable. The decision variables control the change in $S$ through the stage and also determine the return from a stage. The calculus of variations seeks a function that optimizes an *integral*, while dynamic programming seeks to optimize a summation, denoted here as $\sum_{i=1}^{n} r_i$. In the calculus of variations the terminal points of the function $y$ are specified. In dynamic programming, also, $S_n$ and $S'_1$ are specified.

Often some insight is needed to recognize that a physical problem fits into the mold of dynamic programming. Some hint is provided when the problem involves sequences of stages, such as a chain of heat exchangers, reactors, compressors, etc.

So far only the nature of the problem has been described. The next section shows how dynamic programming solves the problem.

## 10-3 CHARACTERISTICS OF THE DYNAMIC-PROGRAMMING SOLUTION

The trademark of dynamic programming in arriving at an overall optimal plan is to establish optimal plans for subsections of the problem. In succeeding evaluations the optimal plans for the subsections are used, and all nonoptimal plans

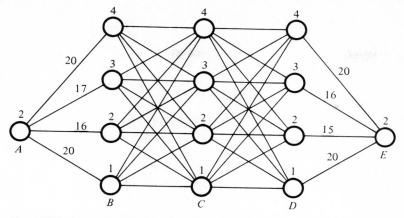

**Figure 10-2** Dynamic programming used to minimize the cost between points $A$ and $E$.

are ignored. The mechanics can be illustrated by a problem of determining the optimal route between two points as in Example 10-1.

**Example 10-1**  A minimum-cost pipeline is to be constructed between points $A$ and $E$, passing successively through one node of each $B$, $C$, and $D$, as shown in Fig. 10-2. The costs from $A$ to $B$ and from $D$ to $E$ are shown in Fig. 10-2, and the costs between $B$ and $C$ and between $C$ and $D$ are given in Table 10-1.

SOLUTION  The strategy in dynamic programming is to begin with one of the terminal stages (either $A$-$B$ or $D$-$E$ in this problem) and then progressively analyze cumulative sets of stages. In this problem it makes no difference whether we start with $A$-$B$ or with $D$-$E$. In some problems the form of the existing data will make starting at one end or the other more convenient.

In this problem we shall start at the right end and work toward the left. Thus the first table will be for stage $D$-$E$, the next table for the cumulative stages $C$-$D$ and $D$-$E$, and so on until the last table applies to the entire system, $A$-$E$.

**Table 10-1  Costs from $B$ to $C$ and $C$ to $D$ in Fig. 10-2**

| From | To | | | |
|---|---|---|---|---|
| | 1 | 2 | 3 | 4 |
| 1 | 12 | 15 | 21 | 28 |
| 2 | 15 | 16 | 17 | 24 |
| 3 | 21 | 17 | 16 | 15 |
| 4 | 28 | 24 | 15 | 12 |

**Table 10-2 Example 10-1 $D$ to $E$**

| From | Through | Cost | Optimum |
|------|---------|------|---------|
| $D1$ | —       | 20   | x       |
| $D2$ | —       | 15   | x       |
| $D3$ | —       | 16   | x       |
| $D4$ | —       | 20   | x       |

Table 10-2 simply shows the costs from the various $D$ positions to the end, which is at $E$. Table 10-2 assumes the form that will be used in subsequent tables; i.e., it designates the optimum manner in which to pass from $D1, D2, D3$, and $D4$ to the next stage, which here is $E$.

Table 10-2 is trivial because there is only one way to pass from $D1$ to the terminus, and that route must be optimum. Future tables will rule out nonoptimal paths, but no such selection can be made in Table 10-2 because it is invalid to say, for example, that the route from $D2$ is less costly than from $D1$, which would permit ruling out $D1$. The consequences of the choice of the routes from $D2$ over that from $D1$ may impose overriding penalties later, so all must be kept in consideration.

It is in Table 10-3 from $C$ to $E$ that dynamic programming begins to

**Table 10-3 Example 10-1, $C$ to $E$**

| From | Through | Cost C to $D$ | Cost D to $E$ | Cost Total | Optimum |
|------|---------|---------------|---------------|------------|---------|
| $C4$ | $D4$    | 12            | 20            | 32         |         |
|      | $D3$    | 15            | 16            | 31         | x       |
|      | $D2$    | 24            | 15            | 39         |         |
|      | $D1$    | 28            | 20            | 48         |         |
| $C3$ | $D4$    | 15            | 20            | 35         |         |
|      | $D3$    | 16            | 16            | 32         | x       |
|      | $D2$    | 17            | 15            | 32         | x       |
|      | $D1$    | 21            | 20            | 41         |         |
| $C2$ | $D4$    | 24            | 20            | 44         |         |
|      | $D3$    | 17            | 16            | 33         |         |
|      | $D2$    | 16            | 15            | 31         | x       |
|      | $D1$    | 15            | 20            | 35         |         |
| $C1$ | $D4$    | 28            | 20            | 48         |         |
|      | $D3$    | 21            | 16            | 37         |         |
|      | $D2$    | 15            | 15            | 30         | x       |
|      | $D1$    | 12            | 20            | 32         |         |

show its benefit. From position $C1$, for example, there are four paths by which to reach $E$, through $D1$, $D2$, $D3$, and $D4$. The total costs are shown in the appropriate column in Table 10-3, and the least costly route is the one passing through $D2$, resulting in a cost of 30. Within that block there is a common basis of comparison: all the paths start from $C1$. Table 10-3 also denotes the optimal choices when starting at the other $C$ positions. In the path from $C3$ there is a tie (cost of 32), so either choice may be made. Now that the preferred route from $C1$ to $E$ has been decided, all the nonoptimal routes are henceforth ignored.

Advancing (backward) to include stage $B$-$C$ in the assembly, Table 10-4 shows the accumulation from $B$ to $E$. The costs are shown for the possible paths starting at all of the $B$ points. From $B1$ it is possible to pass through $C1$ and then to the end, through $C2$ and then to the end, etc. The total costs from $B1$ through $C1$ to the end, for example, are computed by summing the cost of 12 from $B1$ to $C1$ (from Table 10-1) with the cost from $C1$ to the end. The cost of $C1$ to the end is 30, which is available from Table 10-3 as the minimum cost of the four possibilities in the $C1$ block. Of the various paths from $B1$ to the end, one is optimum, with the cost of 42. Table 10-4 shows four state variables, $B1$, $B2$, $B3$, and $B4$, and the optimum cost from each of these states to the end.

The final step in the solution is provided by picking up the first state ($A$ to $B$) in the accumulation, and the result is Table 10-5. There is only one state

**Table 10-4  Example 10-1, $B$ to $E$**

| From | Through | Cost B to C | Cost C to E | Total | Optimum |
|------|---------|------|------|-------|---------|
| $B4$ | $C4$ | 12 | 31 | 43 | x |
|      | $C3$ | 15 | 32 | 47 |   |
|      | $C2$ | 24 | 31 | 55 |   |
|      | $C1$ | 28 | 30 | 58 |   |
| $B3$ | $C4$ | 15 | 31 | 46 | x |
|      | $C3$ | 16 | 32 | 48 |   |
|      | $C2$ | 17 | 31 | 48 |   |
|      | $C1$ | 21 | 30 | 51 |   |
| $B2$ | $C4$ | 24 | 31 | 55 |   |
|      | $C3$ | 17 | 32 | 49 |   |
|      | $C2$ | 16 | 31 | 47 |   |
|      | $C1$ | 15 | 30 | 45 | x |
| $B1$ | $C4$ | 28 | 31 | 59 |   |
|      | $C3$ | 21 | 32 | 53 |   |
|      | $C2$ | 15 | 31 | 46 |   |
|      | $C1$ | 12 | 30 | 42 | x |

**Table 10-5 Example 10-1, $A$ to $E$**

| To $E$ from | Through | Cost | | | Optimum |
| | | $A$ to $B$ | $B$ to $E$ | Total | |
|---|---|---|---|---|---|
| $A2$ | $B4$ | 20 | 43 | 63 | |
| | $B3$ | 17 | 46 | 63 | |
| | $B2$ | 16 | 45 | 61 | x |
| | $B1$ | 20 | 42 | 62 | |

variable, the position $A2$. From $A2$ the options are to pass through $B1, B2$, $B3$, or $B4$. The cost from $A2$ to $B$ is found in Fig. 10-2 and the cost from $B$ to $E2$ is the optimal one from the appropriate $B$ position from Table 10-4.

The minimum cost from $A2$ to $E2$, thus the minimum total cost, is 61. To identify which route results in this minimum cost, it is only necessary to trace back through the tables and note which choices were optimum: $A2$, $B2, C1, D2, E2$.

The key feature of dynamic programming is that after an optimal policy has been determined from an intermediate state to the final state, future calculations passing through that state use only the optimal policy.

## 10-4 EFFICIENCY OF DYNAMIC PROGRAMMING

The combination of tables necessary to solve Example 10-1 may seem to constitute a lengthy and tedious solution. Comparatively speaking, however, dynamic programming is efficient, particularly in large problems. If we call each line in Tables 10-2 to 10-5 a *calculation*, a total of 40 calculations was required. If an exhaustive examination of all possible routes between $A$ and $B$ had been made, the total number would have been the product of the number of possibilities. From $A2$ to $B$ there are four possibilities, from each of the $B$ points there are four possibilities of passing to $C$, and similarly from $C$ to $D$. From $D$ to $E2$ there is just one possibility. The number of possible routes if all are considered is therefore $(4)(4)(4)(1) = 64$.

The saving of effort would be more impressive if the problem had included another stage consisting of four positions. The number of calculations by dynamic programming would have been the current number of 40 plus an additional 16, for a total of 56. Examining all possible routes would require $(64)(4) = 256$ calculations.

**Example 10-2** Whey is a by-product of cheese manufacture and contains, among other constituents, protein and lactose. The two substances are

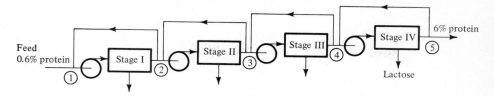

**Figure 10-3** Chain of ultrafilters to separate protein from lactose in whey.

valuable when separated, protein-rich whey for yogurt and lactose for ethyl alcohol. One method of separation is ultrafiltration, where the separation occurs on the basis of molecular size and shape. A series of ultrafilters is employed (four in the chain shown in Fig. 10-3) to separate the protein and lactose progressively. Klinkowski[1] shows the operating cost of a stage to be a function of the inlet and outlet protein concentrations, as shown in Table 10-6. Use dynamic programming to solve for the concentrations leaving each stage that results in the minimum total cost.

SOLUTION The dynamic-programming calculations start, arbitrarily, with stage IV in Table 10-7 and proceed back through the stages until Table 10-10, which includes all the stages. The minimum operating cost is $34.42, which is achieved by operating the filtration plant with concentrations 0.6, 0.9, 1.8, 3.6 to 6.0.

**Table 10-6  Operating cost of one stage in a protein-lactose separator in Example 10-2, dollars**

| Entering protein concentration, % | Leaving protein concentration, % | | | | | | | | | |
|---|---|---|---|---|---|---|---|---|---|---|
| | 0.9 | 1.2 | 1.8 | 2.4 | 3.0 | 3.6 | 4.2 | 4.8 | 5.4 | 6.0 |
| 0.6 | 5.53 | 10.77 | 20.24 | 28.38 | 35.20 | 40.70 | 44.88 | 47.74 | 49.28 | 49.50 |
| 0.9 | | 3.73 | 10.77 | 17.23 | 23.10 | 28.38 | 33.07 | 37.18 | 40.70 | 43.63 |
| 1.2 | | | 5.54 | 10.78 | 15.67 | 20.24 | 24.47 | 28.38 | 31.95 | 35.20 |
| 1.8 | | | | 3.74 | 7.33 | 10.79 | 14.00 | 17.23 | 20.24 | 23.10 |
| 2.4 | | | | | 2.82 | 5.55 | 8.21 | 10.80 | 13.27 | 15.67 |
| 3.0 | | | | | | 2.26 | 4.47 | 6.63 | 8.73 | 10.81 |
| 3.6 | | | | | | | 1.89 | 3.75 | 5.56 | 7.33 |
| 4.2 | | | | | | | | 1.62 | 3.21 | 4.78 |
| 4.8 | | | | | | | | | 1.42 | 2.82 |
| 5.4 | | | | | | | | | | 1.26 |
| 6.0 | | | | | | | | | | |

**Table 10-7  Example 10-2, stage IV**

| Concentration entering stage IV | Through | Cost |
|---|---|---|
| 1.8 | – | $23.10 |
| 2.4 | – | 15.67 |
| 3.0 | – | 10.81 |
| 3.6 | – | 7.33 |
| 4.2 | – | 4.78 |
| 4.8 | – | 2.82 |
| 5.4 | – | 1.26 |

**Table 10-8  Example 10-2, stages III and IV**

| Concentration entering III | Through | Cost | | |
|---|---|---|---|---|
| 1.2 | 1.8 | 5.54 + 23.10 = | $28.64 | |
| | 2.4 | 10.78 + 15.67 = | 26.45* | |
| | 3.0 | 15.67 + 10.81 = | 26.48 | |
| | 3.6 | 20.24 + 7.33 = | 27.57 | |
| | 4.2 | 24.47 + 4.78 = | 29.25 | |
| | 4.8 | 28.38 + 2.82 = | 31.20 | |
| | 5.4 | 31.95 + 1.26 = | 33.21 | |
| 1.8 | 2.4 | 3.74 + 15.67 = | 19.41 | |
| | 3.0 | 7.33 + 10.81 = | 18.14 | |
| | 3.6 | 10.79 + 7.33 = | 18.12* | |
| | 4.2 | 14.00 + 4.78 = | 18.78 | |
| | 4.8 | 17.23 + 2.82 = | 20.05 | |
| | 5.4 | 20.24 + 1.26 = | 21.50 | |
| 2.4 | 3.0 | 2.82 + 10.81 = | 13.63 | |
| | 3.6 | 5.55 + 7.33 = | 12.88* | |
| | 4.2 | 8.21 + 4.78 = | 12.99 | |
| | 4.8 | 10.80 + 2.82 = | 13.62 | |
| | 5.4 | 13.27 + 1.26 = | 14.53 | |
| 3.0 | 3.6 | 2.26 + 7.33 = | 9.59 | |
| | 4.2 | 4.47 + 4.78 = | 9.25* | |
| | 4.8 | 6.63 + 2.82 = | 9.45 | |
| | 5.4 | 8.73 + 1.26 = | 9.99 | |
| 3.6 | 4.2 | 1.89 + 4.78 = | 6.67 | |
| | 4.8 | 3.75 + 2.82 = | 6.57* | |
| | 5.4 | 5.56 + 1.26 = | 6.82 | |
| 4.2 | 4.8 | 1.62 + 2.82 = | 4.44* | |
| | 5.4 | 3.21 + 1.26 = | 4.47 | |
| 4.8 | 5.4 | 1.42 + 1.26 = | 2.68* | |

**Table 10-9  Example 10-2, stages II, III, and IV**

| Concentration entering II | Through | Cost |
|---|---|---|
| 0.9 | 1.2 | 3.73 + 26.45 = $30.18 |
|  | 1.8 | 10.77 + 18.12 = 28.89* |
|  | 2.4 | 17.23 + 12.88 = 30.11 |
|  | 3.0 | 23.10 + 9.25 = 32.35 |
|  | 3.6 | 28.38 + 6.57 = 34.95 |
|  | 4.2 | 33.07 + 4.44 = 37.51 |
|  | 4.8 | 37.18 + 2.68 = 39.86 |
| 1.2 | 1.8 | 5.54 + 18.12 = 23.66* |
|  | 2.4 | 10.78 + 12.88 = 23.66* |
|  | 3.0 | 15.67 + 9.25 = 24.92 |
|  | 3.6 | 20.24 + 6.57 = 26.81 |
|  | 4.2 | 24.47 + 4.44 = 28.91 |
|  | 4.8 | 28.38 + 2.68 = 31.06 |
| 1.8 | 2.4 | 3.74 + 12.88 = 16.62 |
|  | 3.0 | 7.33 + 9.25 = 16.58* |
|  | 3.6 | 10.79 + 6.57 = 17.36 |
|  | 4.2 | 14.00 + 4.44 = 18.44 |
|  | 4.8 | 17.23 + 2.68 = 19.91 |
| 2.4 | 3.0 | 2.82 + 9.25 = 12.07* |
|  | 3.6 | 5.55 + 6.57 = 12.12 |
|  | 4.2 | 8.21 + 4.44 = 12.65 |
|  | 4.8 | 10.80 + 2.68 = 13.48 |
| 3.0 | 3.6 | 2.26 + 6.57 = 8.83* |
|  | 4.2 | 4.47 + 4.44 = 8.91 |
|  | 4.8 | 6.63 + 2.68 = 9.31 |
| 3.6 | 4.2 | 1.89 + 4.44 = 6.33* |
|  | 4.8 | 3.75 + 2.68 = 6.43 |
| 4.2 | 4.8 | 1.62 + 2.68 = 4.30* |

**Table 10-10  Example 10-2, stages I to IV**

| Concentration entering I | Through | Cost |
|---|---|---|
| 0.6 | 0.9 | 5.53 + 28.89 = $34.42* |
|  | 1.2 | 10.77 + 23.66 = 34.43 |
|  | 1.8 | 20.24 + 16.58 = 36.82 |
|  | 2.4 | 28.38 + 12.07 = 40.45 |
|  | 3.0 | 35.20 + 8.83 = 44.03 |
|  | 3.6 | 40.70 + 6.33 = 47.03 |
|  | 4.2 | 44.88 + 4.30 = 49.18 |

## 10-5 THE PATTERN OF THE DYNAMIC-PROGRAMMING SOLUTION

Examples 10-1 and 10-2 begin to display a pattern of the nature of the problem and the characteristics of the dynamic-programming solution. The solutions can be shown graphically as in Fig. 10-4a and b. Example 10-1 is a geometric problem, and the ordinate is a position $y$. The ordinate in Example 10-2 is a physical variable, the protein concentration. The abscissas are the stages; $A$-$B$ is the first stage in Example 10-1, and the first stage in Example 10-2 lies between positions 1 and 2.

Another characteristic of the problems is that the terminal points are specified ($A2$ and $E2$ in Fig. 10-4a and 0.6 and 6 percent in Fig. 10-4b). An optimal path is sought that minimizes a summation which in both problems is a cost.

The first table in the dynamic-programming solution is a necessary routine, but it is in the second table that dynamic programming becomes effective. In Example 10-2 from a given state variable entering stage III, e.g., a concentration of 3.6 percent, the various possible routes to 6 percent at position 5 are investigated. Once the optimum route from 3.6 percent at position 3 to 6 percent at position 5 has been identified, the nonoptimal routes are discarded. Any future path that passes through position 3 at 3.6 percent uses only the optimal route.

The choice of the increments of protein concentration in Example 10-2 (0.6, 0.9, 1.2, 1.8, etc.) was somewhat arbitrary and introduced an approximation. More precision would have resulted if 0.1 percent increments had been chosen and the performance data that yielded Table 10-6 used to compute a more detailed table. The coarser grid was chosen simply to lessen the number of calculations. After having established the approximate optimal path a recalculation could have been made using a finer grid but considering paths only in the neighborhood of the preliminary optimum.

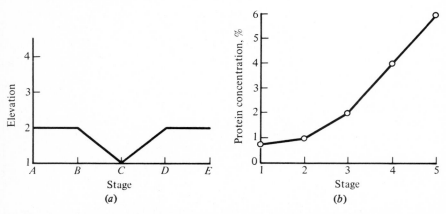

**Figure 10-4** Graphic display of dynamic-programming problems: (a) Example 10-1; (b) Example 10-2.

**Table 10-11  First table of Example 10-2 proceeding forward, stage I**

| Leaving stage I | Entering stage I | Cost, $ |
|---|---|---|
| 0.9 | 0.6 | 5.53 |
| 1.2 | 0.6 | 10.77 |
| 1.8 | 0.6 | 20.24 |
| 2.4 | 0.6 | 28.38 |
| etc. | | |

**Table 10-12  Second table of Example 10-2 proceeding forward, stages I and II**

| Leaving stage II | Through ___, at position 2 | Cost, $ |
|---|---|---|
| 1.2 | 0.9 | $3.73 + 5.53 = 9.26$ |
| 1.8 | 0.9 | $5.53 + 10.77 = 16.30*$ |
|  | 1.2 | $10.77 + 5.54 = 16.31$ |
| 2.4 | 0.9 | $5.53 + 17.23 = 22.76$ |
|  | 1.2 | $10.77 + 10.78 = 21.55*$ |
|  | 1.8 | $20.24 + 3.74 = 23.98$ |
| etc. | | |

It has already been pointed out that the form of the available data may in some problems make starting at the front preferable and in other cases starting at the end and working backward. In Examples 10-1 and 10-2 it was immaterial which direction was chosen. It should be realized, however, that the form of the tables will differ depending upon whether the progression is forward or backward, because the state variable will be different. If the progression in Example 10-2 were forward, the first table would have the form shown in Table 10-11 and the second table would be as shown in Table 10-12.

The state variable in Table 10-12 is the protein percentage leaving stage II, and once it has been determined that the optimal way to achieve a protein concentration of 2.4, for example, through the first two stages (0.6 to 1.2 in the first stage and 1.2 to 2.4 in the second stage, as indicated by Table 10-12), the nonoptimal routes are neglected.

## 10-6  APPARENTLY CONSTRAINED PROBLEMS

An important class of problems in dynamic programming is that of constrained optimization, where a function $y(x)$ is sought that minimizes a summation $\Sigma g(y, x)$

but in addition some other summation is specified $\Sigma h(y, x) = H$, where the functions $g$, $h$, and the numerical term $H$ are known. This class of constrained problems will not be treated in this book. Another class of problems at first glance may seem to be constrained, but they can be converted into a form identical to that used in Examples 10-1 and 10-2. This class may be called *apparently constrained* and is illustrated by Examples 10-3 and 10-4.

**Example 10-3** An evaporator which boils liquid inside tubes consists of four banks of tubes. Each bank consists of a number of tubes in parallel, and the banks are connected in series, as shown in Fig. 10-5. A mixture of liquid and vapor enters the first bank with a fraction of vapor $x = 0.2$, and the fluid leaves the evaporator as saturated vapor, $x = 1.0$. The flow rate is 0.5 kg/s, and each tube is capable of vaporizing 0.01 kg/s and thus of increasing $x$ by 0.02.

Forty tubes are to be arranged in the banks so that the minimum total pressure drop prevails in the evaporator. The pressure drop in a bank is approximately proportional to the square of the velocity, and a satisfactory expression for the pressure drop $\Delta p$ is

$$\Delta p, \text{kPa} = 720 \left( \frac{x_i}{n} \right)^2 \tag{10-1}$$

where $x_i$ = vapor fraction entering bank
$\quad n$ = number of tubes in bank

Use dynamic programming to determine the distribution of the 40 tubes so that the total pressure drop in the evaporator is minimum.

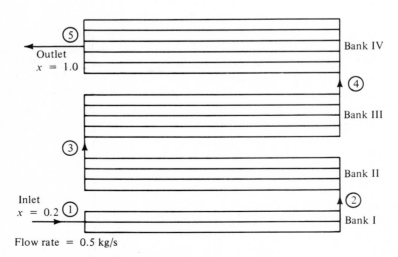

Figure 10-5 Evaporator in Example 10-3.

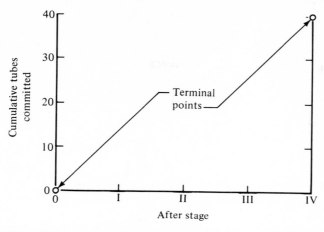

**Figure 10-6** State variable of cumulative number of tubes committed in Example 10-3.

SOLUTION  Selection of the number of tubes in a stage (bank) as the state variable is unproductive because the terminal points in graphs comparable to Fig. 10-4 are not fixed. Furthermore, the choice of number of tubes as the state variable does not account for the constraint that precisely 40 tubes are available.

The difficulties are overcome by choosing as the state variable cumulative tubes committed, which results in coordinates as shown in Fig. 10-6. After stage 0 (before stage I) no tubes have been committed, and following stage IV 40 tubes have been committed.

Table 10-13 shows pressure drops for the first bank or stage for several different choices of tubes. Table 10-14 uses as the state variable the total number of tubes committed in the first two stages and permits an optimal selection. For example, if 13 tubes are used in the first two stages, the optimal distribution is to allot 5 tubes in bank I and 8 tubes in bank II to achieve a total $\Delta p$ in banks I and II of 2.16 kPa. The pressure drop in stage II is computed from Eq. (10-1) using $x_i = 0.20 + (0.02)$ (number of tubes in stage I).

**Table 10-13  Example 10-3, stage I**

| Total tubes committed | Tubes in stage I | Total $\Delta p$, kPa |
|---|---|---|
| 2 | 2 | 7.20 |
| 3 | 3 | 3.20 |
| 4 | 4 | 1.80 |
| 5 | 5 | 1.15 |
| 6 | 6 | 0.80 |

**Table 10-14  Example 10-3, stages I and II**

| Total tubes | Tubes in II | Total $\Delta p$, kPa |
|---|---|---|
| 11 | 5 | 0.80 + 2.95 = 3.75 |
|  | 6 | 1.15 + 1.80 = 2.95* |
|  | 7 | 1.80 + 1.15 = 2.95* |
|  | 8 | 3.20 + 0.76 = 3.96 |
|  | 9 | 7.20 + 0.51 = 7.70 |
| 12 | 6 | 0.80 + 2.05 = 2.85 |
|  | 7 | 1.15 + 1.32 = 2.47* |
|  | 8 | 1.80 + 0.88 = 2.68 |
|  | 9 | 3.20 + 0.60 = 3.68 |
| 13 | 7 | 0.80 + 1.50 = 2.30 |
|  | 8 | 1.15 + 1.01 = 2.16* |
|  | 9 | 1.80 + 0.73 = 2.53 |
|  | 10 | 3.20 + 0.49 = 3.69 |
| 14 | 7 | 0.59 + 1.70 = 2.24 |
|  | 8 | 0.80 + 1.15 = 1.95* |
|  | 9 | 1.15 + 0.80 = 1.95* |
|  | 10 | 1.80 + 0.56 = 2.36 |
| 15 | 8 | 0.59 + 1.30 = 1.89 |
|  | 9 | 0.80 + 0.91 = 1.71* |
|  | 10 | 1.15 + 0.65 = 1.80 |

**Table 10-15  Example 10-3, stages I to III**

| Total tubes | Tubes in III | Total $\Delta p$, kPa |
|---|---|---|
| 22 | 9 | 2.16 + 1.88 = 4.04 |
|  | 10 | 2.47 + 1.39 = 3.86* |
|  | 11 | 2.95 + 1.05 = 4.00 |
| 23 | 9 | 1.95 + 2.05 = 4.00 |
|  | 10 | 2.16 + 1.52 = 3.68 |
|  | 11 | 2.47 + 1.15 = 3.62* |
|  | 12 | 2.95 + 0.88 = 3.83 |
| 24 | 10 | 1.95 + 1.66 = 3.61 |
|  | 11 | 2.16 + 1.26 = 3.42* |
|  | 12 | 2.47 + 0.97 = 3.44 |
|  | 13 | 2.95 + 0.75 = 3.70 |
| 25 | 10 | 1.71 + 1.80 = 3.51 |
|  | 11 | 1.95 + 1.37 = 3.32 |
|  | 12 | 2.16 + 1.06 = 3.22* |
|  | 13 | 2.47 + 0.82 = 3.29 |
| 26 | 11 | 1.71 + 1.49 = 3.20 |
|  | 12 | 1.95 + 1.15 = 3.00* |
|  | 13 | 2.16 + 0.90 = 3.06 |

**Table 10-16  Example 10-3, stages I to IV**

| Total tubes | Tubes in IV | Total $\Delta p$, kPa |
|---|---|---|
| 40 | 13 | 2.93 + 2.33 = 5.26 |
|    | 14 | 3.00 + 1.90 = 4.90 |
|    | 15 | 3.22 + 1.57 = 4.79 |
|    | 16 | 3.42 + 1.30 = 4.72 |
|    | 17 | 3.62 + 1.09 = 4.71* |
|    | 18 | 3.86 + 0.91 = 4.77 |

The optimal distribution of tubes is 5, 7, 11, 17, resulting in a total pressure drop of 4.71 kPa.

A further illustration of the solution by dynamic programming of an apparently constrained problem is shown in Example 10-4 in the optimization of feedwater heating. Heating the boiler feedwater with extraction steam, as shown in Fig. 10-7, improves the efficiency of a steam-power cycle and is a common practice in large central power stations. Some plants use more than half a dozen heaters, which draw off extraction steam at as many different pressures. That feedwater heaters improve the efficiency of a cycle can be shown by a calculation of a specific case, but a qualitative explanation may provide a better sense of this improvement. First, recall that in the steam-power cycle approximately 3 J of heat is supplied at the boiler for every joule of work at the turbine shaft.

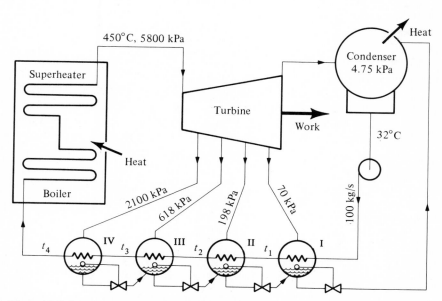

**Figure 10-7** Selection of optimum areas of feedwater heaters in Example 10-4.

The difference of 2 J is the amount rejected at the condenser, which usually represents a loss. The proposal to try to use some of the heat rejected at the condenser for boiler-water heating is doomed because if we tried, for example, to heat the feedwater with exhaust steam from the turbine, there would be no temperature difference between the exhaust steam and the feedwater to provide the driving force for heat transfer.

Extraction steam, however, has a higher temperature than exhaust steam and can be used for the heating. Concentrating on 1 kg of extraction steam leaving the boiler, we find that it performs some work in the turbine before extraction and then uses the remainder of its energy above saturated liquid at the condensing temperature to heat the feedwater. In effect, then, all the heat supplied to that kilogram of steam in the boiler is eventually converted into work. The practice of feedwater heating by extraction steam raises the effectiveness of the cycle compared to rejecting 2 J of boiler heat per joule of work.

It is further to be expected that the high-pressure steam is more valuable than the low-pressure steam because the steam extracted at high pressure would have been able to deliver additional work at the turbine shaft.

**Example 10-4** An economic analysis has determined that a total of 1000 $m^2$ of heat-transfer area should be used in the four feedwater heaters shown in Fig. 10-7. This 1000 $m^2$ can be distributed in the four heaters in 100-$m^2$ increments. The overall heat-transfer coefficient of all heaters is 2800 W/(m² · K). The cost of heat at the boiler is 60 cents per gigajoule, and the worth of the extraction steam determined by thermodynamic calculations is listed in Table 10-17. The flow rate of feedwater is 100 kg/s.

Use dynamic programming to determine the optimum distribution of the area.

SOLUTION Before beginning the solution it may be instructive to try to predict the nature of the optimal solution. It is desirable to use the lowest-cost steam possible, which would suggest a large area in stage I, but each additional unit of area in that stage is less effective than the previous unit area because the temperature difference between the steam and feedwater is less. Thus there must be a compromise between trying to use the low-cost steam and maintaining a high temperature difference.

**Table 10-17 Extraction-steam data**

| Extraction point and heater number | Saturation temperature, °C | Worth of extraction steam, cents per gigajoule |
|---|---|---|
| 1 | 90 | 23 |
| 2 | 120 | 28 |
| 3 | 160 | 38 |
| 4 | 215 | 47 |

**Table 10-18  Example 10-4, stage I**

| Total area committed, $m^2$ | Area for stage I | $t_1$ | Saving per second |
|---|---|---|---|
| 0 | 0 | 32.00 | $.000 |
| 100 | 100 | 60.27 | .438 |
| 200 | 200 | 74.76 | .663 |
| 300 | 300 | 82.19 | .778 |
| 400 | 400 | 86.00 | .837 |
| 500 | 500 | 87.95 | .867 |
| 600 | 600 | 88.95 | .883 |
| 700 | 700 | 89.46 | .891 |
| 800 | 800 | 89.72 | .895 |
| 900 | 900 | 89.86 | .897 |
| 1000 | 1000 | 89.93 | .898 |

This problem is apparently constrained because the total area is specified, but it can be converted into the unconstrained form by using as the state variable the *total area committed*. In this problem it is advantageous to start at the front (with respect to the feedwater flow), because the inlet temperature of the feedwater is known there (32°C). Table 10-18 shows the outlet temperatures from stage I for various areas in that stage. As is typical of dynamic programming, this first table is routine. The temperatures are

**Table 10-19  Example 10-4, stages I and II**

| Total area | Area in II | $t_2$ | Savings per second |
|---|---|---|---|
| 0 | 0 | 32.00 | $  .000 |
| 100 | 0 | 60.27 | .438 |
|  | 100 | 74.89 | .575* |
| 200 | 0 | 74.76 | .663 |
|  | 100 | 89.38 | .829 |
|  | 200 | 96.88 | .870* |
| . . . . . . . . . . . . . . . . . . . . . . . . . . . . . . . . . . . . |  |  |  |
| 1000 | 0 | 89.93 | .898 |
|  | 100 | 104.55 | 1.094 |
|  | 200 | 112.04 | 1.194 |
|  | 300 | 115.89 | 1.245 |
|  | 400 | 117.86 | 1.270 |
|  | 500 | 118.87 | 1.282 |
|  | 600 | 119.38 | 1.285* |
|  | 700 | 119.65 | 1.280 |
|  | 800 | 119.78 | 1.267 |
|  | 900 | 119.85 | 1.237 |
|  | 1000 | 119.89 | 1.178 |

computed by use of Eq. (5-10) for a condenser, and the saving is the value of the heat saved at the boiler less the cost of extraction steam used.

Table 10-19 uses as the state variable the total area committed in the first two stages. If, for example, 1000 m² is available for the first two stages, the optimum distribution is to allot 400 m² in the first stage and 600 m² in the second, resulting in a saving of \$1.285 per second.

Table 10-20 shows various area distributions in the first three stages.

Finally, Table 10-21 where the full area of 1000 m² is committed, indi-

**Table 10-20  Example 10-4, stages I to III**

| Total area | Area in III | $t_3$ | Savings per second |
|---|---|---|---|
| 0 | 0 | 32.00 | \$ .000 |
| 100 | 0 | 74.89 | .575* |
|  | 100 | 94.39 | .575* |
| 200 | 0 | 96.88 | .870 |
|  | 100 | 116.37 | .957* |
|  | 200 | 126.37 | .870 |
| . . . . . . . . . . . . . . . . . . . . . . . . . . . . . . . . . . . . . . . . . |
| 1000 | 0 | 119.38 | 1.285 |
|  | 100 | 138.88 | 1.462 |
|  | 200 | 149.14 | 1.548 |
|  | 300 | 154.26 | 1.590 |
|  | 400 | 157.02 | 1.598* |
|  | 500 | 158.37 | 1.597 |
|  | 600 | 159.13 | 1.566 |
|  | 700 | 159.48 | 1.537 |
|  | 800 | 159.70 | 1.449 |
|  | 900 | 159.79 | 1.358 |
|  | 1000 | 159.84 | 1.178 |

**Table 10-21  Example 10-4, stages I to IV**

| Total area | Area stage IV | $t_4$ | Savings per second |
|---|---|---|---|
| 1000 | 0 | 157.02 | \$1.598 |
|  | 100 | 185.18 | 1.738 |
|  | 200 | 198.92 | 1.801 |
|  | 300 | 206.72 | 1.804* |
|  | 400 | 210.68 | 1.790 |
|  | 500 | 212.54 | 1.773 |
|  | 600 | 213.70 | 1.682 |
|  | 700 | 214.19 | 1.625 |
|  | 800 | 214.53 | 1.492 |
|  | 900 | 214.66 | 1.336 |
|  | 1000 | 214.77 | .996 |

cates that the optimum distribution of area is 100, 300, 300, 300 for a total saving of $1.804 per second.

## 10-7 SUMMARY

When optimizing a system that consists of a chain of events or components where the output condition from one unit forms the input to the next, dynamic programming should be explored. On large problems the amount of calculation may be extensive even though it represents only a fraction of the effort of conducting an exhaustive exploration. The systematic nature of dynamic programming lends itself to development of a computer program to perform the calculations. The major challenge usually appears in setting up the tables, especially identifying the state variable.

## PROBLEMS

**10-1** Using dynamic programming, determine the flight plan for a commercial airliner flying between two cities 1200 km apart so that the minimum amount of fuel is consumed during the flight. Specifically, the altitudes at locations $B$ through $F$ in Fig. 10-8 during the course of the flight are to be specified. Table 10-22 shows the fuel consumption for 200-km ground

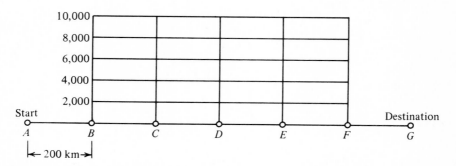

**Figure 10-8** Altitudes and distances in Prob. 10-1.

**Table 10-22 Fuel consumption per 200 km of ground travel, kg**

| From altitude, m | To altitude, m | | | | | |
|---|---|---|---|---|---|---|
| | 0 | 2000 | 4000 | 6000 | 8000 | 10,000 |
| 0 | — | 1500 | 1800 | 2070 | 2300 | 2500 |
| 2,000 | 300 | 600 | 1000 | 1500 | 1950 | 2300 |
| 4,000 | 120 | 180 | 300 | 840 | 1160 | 1730 |
| 6,000 | 0 | 60 | 120 | 200 | 600 | 1200 |
| 8,000 | 0 | 0 | 30 | 80 | 180 | 600 |
| 10,000 | 0 | 0 | 0 | 0 | 60 | 90 |

distances as a function of the climb or descent during that distance. Determine the flight plan and the minimum fuel cost.

**Ans.:** 2770 kg.

**10-2** A minimum-cost pipeline is to be constructed between positions $A$ and $G$ in Fig. 10-9 and can pass through any of six locations in the successive stages, $B, C, D, E$, and $F$.

(a) If all possible combinations of routes are investigated, how many different paths must be examined?

(b) If dynamic programming is used, how many calculations must be made if one calculation consists of one line in a table?

**Ans.:** 7776 and 156.

**10-3** The maintenance schedule for a plant is to be planned so that a maximum total profit will be achieved during a 4-year span that is part of the life of the plant. The income level of the plant during a given year is a function of the condition of the plant carried over from the previous year and the amount of maintenance expenditure at the beginning of the year. Table 10-23 shows the necessary maintenance expenditures that result in a certain income

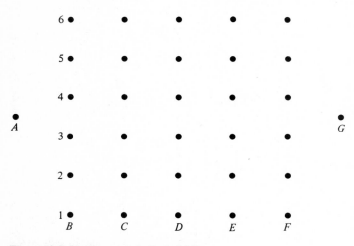

**Figure 10-9** Pipeline route in Prob. 10-2.

**Table 10-23 Maintenance expenditures made at beginning of year, thousands of dollars**

| Income level carried over from previous year | Income level during year | | | | | |
|---|---|---|---|---|---|---|
| | $30 | $32 | $34 | $36 | $38 | $40 |
| $30 | $2 | $4 | $7 | $11 | $16 | $23 |
| 32 | 2 | 3 | 5 | 9 | 13 | 18 |
| 34 | 1 | 2 | 4 | 7 | 10 | 14 |
| 36 | 0 | 1 | 2 | 5 | 8 | 10 |
| 38 | X | 0 | 1 | 2 | 6 | 9 |
| 40 | X | X | 0 | 1 | 4 | 8 |

level during the year for various income levels carried over from the previous year. The income level at the beginning of year 1 just before the maintenance expenses are made is $36,000, and the income level specified during and at the end of year 4 is to be $34,000. The profit for any one year will be the income during the year less the expenditure made for maintenance at the beginning of the year.

Use dynamic programming to determine the plan for maintenance expenditures that results in maximum profit for the 4 years.

**Ans.:** Maximum profit = $130,000.

**10-4** Four heat exchangers (or fewer) in series, as shown in Fig. 10-10, are each served by steam at a different temperature and heat water from 50 to 300°C. The sums of the first costs of the heat exchangers and present worths of the lifetime steam costs are shown in Table 10-24. Use dynamic programming to determine the outlet temperature from each heat exchanger that results in the minimum total present worth of costs.

**Ans.:** $254,300.

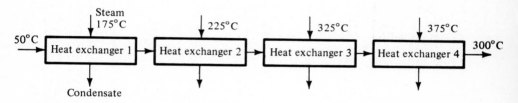

**Figure 10-10** Chain of heat exchangers in Prob. 10-4.

**Table 10-24  Present worth of heat exchanger and lifetime costs of steam, thousands of dollars**

| Heat exchanger | Inlet temp., °C | Outlet temperature, °C | | | | | |
|---|---|---|---|---|---|---|---|
| | | 50 | 100 | 150 | 200 | 250 | 300 |
| 1 | 50 | 0 | $20.8 | $58.0 | X | X | X |
| 2 | 50 | 0 | 23.1 | 62.6 | $132.3 | X | X |
| | 100 | | 0 | 36.1 | 93.6 | X | X |
| | 150 | | | 0 | 62.8 | X | X |
| 3 | 50 | 0 | 24.8 | 79.9 | 129.9 | $177.0 | $308.3 |
| | 100 | | 0 | 41.1 | 94.3 | 141.7 | 266.7 |
| | 150 | | | 0 | 51.2 | 103.0 | 223.6 |
| | 200 | | | | 0 | 57.6 | 176.4 |
| 4 | 50 | | | | | | 372.4 |
| | 100 | | | | | | 309.3 |
| | 150 | | | | | | 243.7 |
| | 200 | | | | | | 173.7 |
| | 250 | | | | | | 94.4 |
| | 300 | | | | | | 0 |

**Table 10-25  Pumping energy**

| Temperature difference at start of 3-h period, $t_w - t_{wb}$† | Drop in $t_w$ in 3-h period, °C | | | | | |
|---|---|---|---|---|---|---|
| | 0 | 0.5 | 1.0 | 1.5 | 2.0 | 2.5 |
| 1.5 | 0 | 2 | 16 | 54 | X | X |
| 2.0 | | 1 | 12 | 45 | X | X |
| 2.5 | | 0 | 8 | 36 | 96 | X |
| 3.0 | | | 4 | 27 | 80 | 166 |
| 3.5 | | | 0 | 18 | 64 | 139 |
| 4.0 | | | | 9 | 48 | 111 |
| 4.5 | | | | 0 | 32 | 83 |
| 5.0 | | | | | 16 | 55 |
| 5.5 | | | | | 0 | 27 |
| 6.0 | | | | | | 0 |

†$t_w$ = temperature of water in pond, °C
$t_{wb}$ = wet-bulb temperature of ambient air, °C

**10-5** A cooling pond serving a power plant is equipped with circulating pumps and sprays to enhance the rate of heat rejection from the pond. Furthermore, the pumps are to be operated so that the heat rejection is accomplished with minimal pumping energy. On one particular day the pond temperature is 28.5°C at 1800 hours, and the pond temperature is to be reduced to 21.5°C by 0600 the next morning. The rate of decrease of temperature of the pond water is a function of the temperature difference between the pond water and the ambient wet-bulb temperature as well as the intensiveness of pumping. The pumping energies during a 3-h period are shown in Table 10-25.

The ambient wet-bulb temperatures at the beginning of each 3-h period are 1800, 25.0°C; 2100, 23.0°C; 2400, 21.5°C; and 0300, 20.0°C. Use dynamic programming to determine the pond-water temperature at each 3-h interval that results in minimum pumping energy.

**Ans.:** Minimum energy = 139.

**10-6** A rocket starting from rest carries an initial fuel charge of 10,000 kg, which is to be burned in 120 s at such a rate that the maximum velocity of the rocket is to be achieved at the end of the burning time. The 120-s burning time is divided into four 30-s intervals. Table 10-26 presents the increase in velocity in each 30-s interval as a function of the mass of fuel in the rocket at the start of the interval and the mass of fuel burned during the interval. All 10,000 kg is to be expended in 120 s, and at least 1000 kg is to be burned each time interval.

Use dynamic programming to determine the fuel-burning plan that results in the highest velocity of the rocket in 120 s.

**Ans.:** 1533 m/s.

**10-7** Hydrazine, $N_2H_4$, is a possible fuel for an emergency-use gas turbine because in the presence of a catalyst it decomposes in an exothermic reaction

$$2N_2H_4 \longrightarrow 2NH_3 + N_2 + H_2$$

The reactor is to consist of four stages, as shown in Fig. 10-11. A particular reactor is limited to a total of 14 kg of catalyst, which is available in 1-kg packages. Each stage can accommodate 0 to 5 kg of catalyst. The fraction of hydrazine undecomposed in a stage is given by

$$\frac{y_o}{y_i} = 0.5 + 0.5e^{-my_i}$$

**Table 10-26  Increase in velocity in 30-s interval, m/s**

| Initial mass of fuel, kg | Fuel burned in 30 s, kg | | | | | | |
|---|---|---|---|---|---|---|---|
| | 1000 | 2000 | 3000 | 4000 | 5000 | 6000 | 7000 |
| 1,000 | 226 | | | | | | |
| 2,000 | 200 | 424 | | | | | |
| 3,000 | 180 | 377 | 599 | | | | |
| 4,000 | 165 | 341 | 536 | 756 | | | |
| 5,000 | 152 | 312 | 486 | 679 | 895 | | |
| 6,000 | 143 | 288 | 445 | 617 | 808 | 1024 | |
| 7,000 | 136 | 269 | 412 | 567 | 737 | 926 | 1140 |
| 8,000 | 131 | 253 | 385 | 525 | 679 | 847 | 1034 |
| 9,000 | 127 | 241 | 362 | 491 | 630 | 781 | 947 |
| 10,000 | 124 | 231 | 343 | 462 | 589 | 726 | 875 |

where $y_i$ = fraction of undecomposed hydrazine entering stage
$y_o$ = fraction of undecomposed hydrazine leaving stage
$m$ = mass of catalyst used in stage

Use dynamic programming to determine the optimum distribution of the 14 kg of catalyst. To shorten the calculation effort, construct the tables only in the neighborhood of the answer given below.

**Ans.:** 2, 3, 4, 5.

**10-8** Some impurities in industrial waste water can be removed[2] by passing the water through a series of adsorbers containing activated carbon. In a certain installation the adsorbers are arranged in four stages with waste water having an initial contamination of 2000 ppm entering the first stage. A total of 140 kg of activated carbon is available to be distributed in the four stages in quantities of 10, 20, 30, 40, or 50 kg in each stage. The performance of each stage is expressed by

$$x_o = x_i \left( 1 - \frac{m}{60} + \frac{m^2 x_i}{20{,}000{,}000} \right)$$

where $x_o$ = contamination at outlet of stage, ppm
$x_i$ = contamination at inlet of stage, ppm
$m$ = mass of activated carbon in the stage, kg

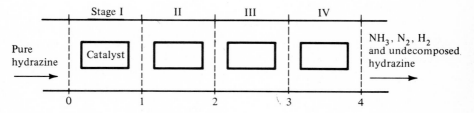

**Figure 10-11** Four-stage hydrazine reactor.

Use dynamic programming to establish the distribution that results in the minimum contamination of the waste water leaving stage 4. To shorten the calculation effort, construct the tables only in the neighborhood of the answer given below.

    **Ans.:** 10, 30, 50, 50.

# REFERENCES

1. P. R. Klinkowski, "Ultrafiltration: An Emerging Unit Operation," *Chem. Eng.*, vol. 85, no. 11, pp. 164–173, May 6, 1978.
2. J. L. Rizzo and A. R. Shepherd, "Treating Industrial Wastewater with Activated Carbon," *Chem. Eng.*, vol. 84, no. 1, pp. 95–100, Jan. 3, 1977.

# ADDITIONAL READINGS

Bellman, R. E.: *Dynamic Programming*, Princeton University Press, Princeton, N.J., 1957.

Bellman, R. E., and S. Dreyfus: *Applied Dynamic Programming*, Princeton University Press, Princeton, N.J., 1962.

Denn, M. M.: *Optimization by Variational Methods*, McGraw-Hill, New York, 1969.

Gluss, B.: *An Elementary Introduction to Dynamic Programming; a State Equation Approach*, Allyn and Bacon, Boston, 1972.

Hastings, N. A. J.: *Dynamic Programming with Management Applications*, Crane, New York, 1973.

Nemhauser, G. L.: *Introduction to Dynamic Programming*, Wiley, New York, 1960.

Roberts, S.: *Dynamic Programming in Chemical Engineering and Process Control*, Academic, New York, 1964.

## 11-1 INTRODUCTION

Geometric programming is one of the newest methods of optimization. Clarence Zener first recognized the significance of the geometric and arithmetic mean in cases of unconstrained optimization. Since then others have extended the methods to accommodate more general optimization problems. The form of problem statement that is particularly adaptable to treatment by geometric programming is a sum of polynomials for both the objective function and the constraint equations. These polynomials can be made up of combinations of variables to either positive or negative noninteger or integer exponents. After seeing the usefulness of polynomial representations in Chap. 4, we recognize that the ability to optimize such functions is clearly of engineering importance.

A feature of geometric programming is that the first stage of the solution is to find the optimum value of the function, rather than first to determine the values of the independent variables that give the optimum. This knowledge of the optimum value may be all that is of interest, and the calculation of the values of the variables can be omitted.

This chapter first presents the form of a geometric programming problem and then defines *degree of difficulty*, because problems with zero degree of difficulty are ideally suited to geometric programming. Furthermore, optimization problems with degree of difficulty greater than zero are probably best solved by some other method of optimization. Next the mechanics of the method will be illustrated with several examples. Gradually this chapter weaves in some explanation and proof of geometric programming so that the user will have confidence in a technique that at first may seem to be a dark art.

223

Unconstrained optimizations are the first ones attacked, and later the study moves to constrained optimization with equality constraints. Geometric programming is capable of optimizing objective functions subject to inequality constraints, but this application is beyond the scope of this introductory chapter.

In addition to providing optimal values of the objective function and independent variables, geometric programming supplies additional insight into the solution. For example, the solution by geometric programming also shows how the total cost is divided among the various contributors.

## 11-2 FORM OF THE OBJECTIVE FUNCTION AND CONSTRAINTS

Geometric programming is adaptable to problems where the objective function and constraints are sums of polynomials of the variables. The variables can be taken to integer or noninteger positive or negative exponents. The following examples of unconstrained objective functions can be solved by geometric programming:

Minimize
$$y = 5x + \frac{10}{\sqrt{x}} \tag{11-1}$$

Maximize
$$y = 6 + 3x - x^{1.8} \tag{11-2}$$

Minimize
$$y = 2x_1 x_2^2 + \frac{5}{\sqrt{x_1 x_2}} + 10x_1^{0.8} - 4x_2 \tag{11-3}$$

An example of a constrained optimization adaptable to geometric programming is

Minimize
$$y = 5x_1\sqrt{x_2} + 2x_1^2 + x_2^{3/2} \tag{11-4}$$

subject to
$$x_1 x_2 = 50$$

## 11-3 DEGREE OF DIFFICULTY

Duffin, Peterson, and Zener[1] define degree of difficulty when applied to geometric programming problems as $T - (N + 1)$, where $T$ is the number of terms in the objective function plus those in the constraints and $N$ is the number of variables. The degree of difficulty of the objective function in Eq. (11-1) is $2 - (1 + 1) = 0$. The degree of difficulty of the objective function in Eq. (11-2) is also zero, since only $3x - x^{1.8}$ will be optimized. The degree of difficulty of Eq. (11-3) is unity. In the constrained optimization of the objective function in Eq. (11-4), the number of terms is $3 + 1 = 4$ (three in the objective function and one in the constraint), and the number of variables is 2, so the degree of difficulty is 1.

When the degree of difficulty is zero, geometric programming is often the simplest method available for solution. For degrees of difficulty greater than zero, geometric programming will work,[2] but the method involves the solution of nonlinear equations, which will probably be more time-consuming than if some other method, such as Lagrange multipliers, is used. Henceforth in this chapter, only problems of zero degree of difficulty will be considered.

## 11-4 MECHANICS OF SOLUTION FOR ONE INDEPENDENT VARIABLE, UNCONSTRAINED

Before providing the support for the method, the mechanics will be presented through several examples. The optimal value $y^*$ will be sought for the function

$$y = c_1 x^{a_1} + c_2 x^{a_2} \tag{11-5}$$

The individual terms will be designated by the symbol $u$; thus

$$u_1 = c_1 x^{a_1} \qquad u_2 = c_2 x^{a_2} \qquad y = u_1 + u_2$$

Geometric programming asserts that the optimal value $y^*$ can also be represented in product form by an expression that we shall call $g^*$,

$$y^* = g^* = \left(\frac{c_1 x^{a_1}}{w_1}\right)^{w_1} \left(\frac{c_2 x^{a_2}}{w_2}\right)^{w_2} \tag{11-6}$$

provided that

$$w_1 + w_2 = 1 \tag{11-7}$$

and

$$a_1 w_1 + a_2 w_2 = 0 \tag{11-8}$$

A consequence of Eq. (11-8) is that the $x$'s cancel out of Eq. (11-6), so the solution is

$$y^* = g^* = \left(\frac{c_1}{w_1}\right)^{w_1} \left(\frac{c_2}{w_2}\right)^{w_2} \tag{11-9}$$

where $w_1$ and $w_2$ are specified by Eqs. (11-7) and (11-8).

A further significance of $w_1$ and $w_2$ is that at the optimum

$$w_1 = \frac{u_1^*}{u_1^* + u_2^*} = \frac{u_1^*}{y^*} \tag{11-10}$$

$$w_2 = \frac{u_2^*}{u_1^* + u_2^*} = \frac{u_2^*}{y^*} \tag{11-11}$$

Equations (11-10) and (11-11) may be useful in solving for $x^*$.

**Example 11-1** Determine the optimum pipe diameter which results in minimum first plus operating cost for 100 m of pipe conveying a given water flow rate. The first cost of the installed pipe in dollars is $160D$, where $D$ is the pipe diameter in millimeters. The lifetime pumping cost is $(32 \times 10^{12})/$

$D^5$ dollars. It is to be expected that the pumping cost will be proportional to $D^{-5}$ because the pumping cost for a specified flow rate $Q$, number of hours of operation, pump efficiency, motor efficiency, and electric rate is proportional to the pressure drop in the pipe. Further, this pressure drop $\Delta p$ is

$$\Delta p, \text{Pa} = f \frac{L}{D} \frac{V^2 \rho}{2} = f \frac{L}{D} \left( \frac{Q}{\pi D^2/4} \right)^2 \frac{\rho}{2} = \frac{\text{const}}{D^5}$$

The objective function, the cost $y$, in terms of the variable $D$ is then

$$y = 160D + \frac{32 \times 10^{12}}{D^5} \tag{11-12}$$

SOLUTION 1 To provide a check on the geometric-programming method, optimize by calculus,

$$\frac{dy}{dD} = 160 - \frac{(5)(32 \times 10^{12})}{D^6} = 0$$

Then $D^* = 100$ mm and $y^* = \$19,200$

Next optimize Eq. (11-12) by geometric programming.

SOLUTION 2

$$y^* = 160D^* + \frac{32 \times 10^{12}}{(D^*)^5} = g^*$$

$$y^* = g^* = \left( \frac{160D}{w_1} \right)^{w_1} \left( \frac{32 \times 10^{12}}{D^5 w_2} \right)^{w_2}$$

provided that $w_1$ and $w_2$ are chosen so that

$$a_1 w_1 + a_2 w_2 = w_1 - 5w_2 = 0$$

and $$w_1 + w_2 = 1$$

Solving gives $w_1 = \frac{5}{6}$ and $w_2 = \frac{1}{6}$

Substituting these values for $w_1$ and $w_2$ into the expression for $g^*$ results in the cancellation of the $D$'s, leaving

$$y^* = g^* = \left( \frac{160}{5/6} \right)^{5/6} \left( \frac{32 \times 10^{12}}{1/6} \right)^{1/6}$$

$$y^* = g^* = \$19,200$$

The value of $D^*$ can be found by applying Eq. (11-10)

$$w_1 = \frac{5}{6} = \frac{u_1^*}{u_1^* + u_2^*} = \frac{u_1^*}{y^*} = \frac{160D^*}{19,200}$$

and so $D^* = 100$ mm = optimum diameter

The following example is another illustration of the mechanics of geometric programming applied to an unconstrained function of one independent variable. It differs from Example 11-1 in that the objective function now contains a negative term.

**Example 11-2** The torque $T$ in newton-meters developed by a certain internal combustion engine is represented by

$$T = 23.6\omega^{0.7} - 3.17\omega$$

where $\omega$ is the rotative speed in radians per second. Determine the maximum power of which this engine is capable and the rotative speed at which the maximum occurs.

SOLUTION The power $P$ in watts is the product of the torque and the rotative speed,

$$P = T\omega = 23.6\omega^{1.7} - 3.17\omega^2$$

Applying geometric programming, we have

$$y^* = g^* = \left(\frac{23.6}{w_1}\right)^{w_1} \left(\frac{-3.17}{w_2}\right)^{w_2} \tag{11-13}$$

provided that

$$w_1 + w_2 = 1 \quad \text{and} \quad 1.7w_1 + 2w_2 = 0$$

Solving the two equations for $w_1$ and $w_2$ gives

$$w_1 = 6.667 \quad \text{and} \quad w_2 = -5.667$$

Substituting these values of $w_1$ and $w_2$ into Eq. (11-13) gives

$$g^* = \left(\frac{23.6}{6.667}\right)^{6.667} \left(\frac{-3.17}{-5.667}\right)^{-5.667}$$

$$= 122{,}970 \text{ W} = 123.0 \text{ kW} = y^*$$

To compute the value of $\omega^*$, use the fact that

$$u_1^* = 23.6\omega^{*1.7} = y^*w_1 = (122{,}970)(6.667)$$

so
$$\omega^* = \left[\frac{(122{,}970)(6.667)}{23.6}\right]^{1/1.7} = 469 \text{ rad/s} = 74.6 \text{ r/s}$$

An alternate means of determining $\omega^*$ would be to use $u_2^*$,

$$u_2^* = (-3.17)(\omega^*)^2 = 122{,}970w_2 = (122{,}970)(-5.667)$$

$$(\omega^*)^2 = 219{,}830 \quad \omega^* = 469 \text{ rad/s}$$

The existence of the negative coefficient in the objective function caused no special difficulty.

A further example will illustrate yet another class of problems and also a situation which at first may seem to make the problem insoluble.

**Example 11-3** Find $y^*$ and $x^*$ for the function

$$y = -4x + x^{1.6}$$

SOLUTION

$$y^* = g^* = \left(\frac{-4}{w_1}\right)^{w_1} \left(\frac{1}{w_2}\right)^{w_2}$$

provided that $\quad w_1 + w_2 = 1 \quad$ and $\quad w_1 + 1.6w_2 = 0$

The values of the $w$'s are $w_1 = 2.667$ and $w_2 = -1.667$, and so

$$y^* = g^* = \left(\frac{-4}{2.667}\right)^{2.667} \left(\frac{1}{-1.667}\right)^{-1.667} \tag{11-14}$$

Each combination of numbers in the parentheses in Eq. (11-14) is negative, and a negative number taken to a noninteger power is a complex number. The difficulty can be resolved, however, by extracting the negative sign from both terms in parentheses,

$$y^* = (-1)^{2.667} \left(\frac{4}{2.667}\right)^{2.667} (-1)^{-1.667} \left(\frac{1}{1.667}\right)^{-1.667}$$

$$= (-1) \left(\frac{4}{2.667}\right)^{2.667} \left(\frac{1}{1.667}\right)^{-1.667} = -6.90$$

The optimal value of $x$ can be found next

$$-4x = u_1^* = w_1 y^* = (2.667)(-6.90)$$

$$x^* = 4.6$$

The optimal value of $y$ in this example is negative, which makes the components complex numbers in Eq. (11-14). It is possible, however, to extract the negative terms and group them as negative unity taken to the power of unity.

## 11-5 WHY GEOMETRIC PROGRAMMING WORKS; ONE INDEPENDENT VARIABLE

The previous section presented the mechanics of optimizing a function of one independent variable by using geometric programming but gave no substantiation for the method. This section will prove the validity of the geometric-programming method for the one-independent-variable problem. The form of the objective function is the same as that of Eq. (11-5)

$$y = c_1 x^{a_1} + c_2 x^{a_2} = u_1 + u_2 \tag{11-15}$$

Next fabricate a function $g$ such that

$$g = \left(\frac{u_1}{w_1}\right)^{w_1} \left(\frac{u_2}{w_2}\right)^{w_2} = \left(\frac{c_1 x^{a_1}}{w_1}\right)^{w_1} \left(\frac{c_2 x^{a_2}}{w_2}\right)^{w_2} \tag{11-16}$$

where

$$w_1 + w_2 = 1 \tag{11-17}$$

A certain combination of values of $w_1$ and $w_2$ will provide a maximum value of $g$. To determine these values of $w_1$ and $w_2$, apply the method of Lagrange multipliers to Eq. (11-16) subject to the constraint of Eq. (11-17). The maximum values of $g$ and $\ln g$ both occur at the same value of the $w$'s; since it is more convenient to optimize $\ln g$:

Maximize
$$\ln g = w_1 (\ln u_1 - \ln w_1) + w_2 (\ln u_2 - \ln w_2) \tag{11-18}$$

subject to
$$w_1 + w_2 - 1 = \phi = 0 \tag{11-19}$$

Use the method of Lagrange multipliers

$$\nabla (\ln g) - \lambda \, \nabla \phi = 0$$

$$\phi = 0$$

which provides the three equations

$w_1 :$
$$\ln u_1 - 1 - \ln w_1 - \lambda = 0$$

$w_2 :$
$$\ln u_2 - 1 - \ln w_2 - \lambda = 0$$

$$w_1 + w_2 - 1 = 0$$

The unknowns are $w_1$, $w_2$, and $\lambda$, and the solutions for $w_1$ and $w_2$ are

$$w_1 = \frac{u_1}{u_1 + u_2} \tag{11-20}$$

and
$$w_2 = \frac{u_2}{u_1 + u_2} \tag{11-21}$$

Substituting these values of $w_1$ and $w_2$ into Eq. (11-16) gives

$$g = \left[\frac{u_1}{u_1/(u_1 + u_2)}\right]^{u_1/(u_1 + u_2)} \left[\frac{u_2}{u_2/(u_1 + u_2)}\right]^{u_2/(u_1 + u_2)}$$

Thus
$$g = u_1 + u_2$$

Let us pause at this point to assess our status. By the choice of $w_1$ and $w_2$ according to Eqs. (11-20) and (11-21) the value of $g$ is made equal to that of $u_1 + u_2$ and therefore also to $y$. Any other combination of $w_1$ and $w_2$ results in a value of $g$ that is less than $u_1 + u_2 = y$.

Since our original objective is to minimize $y$, the next step is to use the value of $x$ in Eq. (11-15) that results in the minimum value of $u_1 + u_2$. The value of $g$ at this condition $g^*$ will therefore be the one where the $w$'s are chosen such that $g$ always equals $y$ but also with the value of $x$ chosen so that $y$ is a minimum. This value of $x^*$ can be found by equating the derivative of Eq. (11-15) to zero

$$a_1 c_1 x^{(a_1-1)} + a_2 c_2 x^{(a_2-1)} = 0$$

Multiplying by $x$ gives

$$a_1 c_1 x^{a_1} + a_2 c_2 x^{a_2} = 0$$

and so

$$a_1 u_1^* + a_2 u_2^* = 0 \qquad (11\text{-}22)$$

where $u_1^*$ and $u_2^*$ are the values of $u_1$ and $u_2$ at the minimum value of $y$.

From Eq. (11-22)

$$u_1^* = -\frac{a_2 u_2^*}{a_1}$$

which, when substituted into Eqs. (11-20) and (11-21), yields

$$w_1 = \frac{-(a_2/a_1)u_2^*}{-(a_2/a_1)u_2^* + u_2^*} = \frac{-a_2}{a_1 - a_2} \qquad (11\text{-}23)$$

and

$$w_2 = \frac{u_2^*}{-(a_2/a_1)u_2^* + u_2^*} = \frac{a_1}{a_1 - a_2}$$

When these values of $w_1$ and $w_2$ are substituted into Eq. (11-16) for the exponents, the expression for $g^*$ results

$$g^* = \left(\frac{c_1 x^{a_1}}{w_1}\right)^{-a_2/(a_1-a_2)} \left(\frac{c_2 x^{a_2}}{w_2}\right)^{a_1/(a_1-a_2)}$$

Of special importance is the fact that $x$ can be canceled out, leaving

$$g^* = \left(\frac{c_1}{w_1}\right)^{-a_2/(a_1-a_2)} \left(\frac{c_2}{w_2}\right)^{a_1/(a_1-a_2)}$$

or

$$g^* = \left(\frac{c_1}{w_1}\right)^{w_1} \left(\frac{c_2}{w_2}\right)^{w_2} = y^* \qquad (11\text{-}24)$$

In executing geometric programming the values of $w$ proportion themselves so that the $x$'s cancel in the expression for $g$.

## 11-6 SOME INSIGHTS PROVIDED BY GEOMETRIC PROGRAMMING

Geometric programming not only yields optimal values of the variables and objective function but can provide some insight into the solution. Consider, for instance, Example 11-1, where the life-cycle cost of the pump-pipe system was optimized:

$$y = \text{pipe cost} + \text{present worth of lifetime energy cost}$$

$$= 160D + \frac{32 \times 10^{12}}{D^5}$$

For this objective function the optimal value of diameter $D^* = 100$ mm and $y^* = \$19,200$. The values of the weighting factors, $w_1$ and $w_2$ of $\frac{5}{6}$ and $\frac{1}{6}$, respectively, are also of interest, because at the optimum five-sixths of the total cost is devoted to the pipe and one-sixth to the energy.

Let us suppose that the cost of the energy increases. What would be the effect on the optimum? What happens to the solution, for example, if the lifetime energy cost becomes $(50 \times 10^{12})/D^5$? Of particular importance is the fact that the distribution of the total cost remains unchanged (five-sixths and one-sixth) because the exponents of $D$ that control $w_1$ and $w_2$ remain unchanged. Both $y^*$ and $D^*$ increase; the optimum total cost now becomes \$20,680, and $D = 107.7$ mm. The increase in optimal diameter probably conforms to our expectation that upon an increase in the energy cost the diameter responds by increasing. But regardless of what happens to the unit cost of energy or pipe, the distribution between the first cost of the pipe and the energy remains constant.

Suppose, however, that an exponent of $D$ changes, e.g., the cost of the pipe increases at a more rapid rate than linearly. If the cost of the pipe is proportional to $D^{1.2}$, for example, the distribution of costs between the pipe and energy changes because the new value of $w_1 = 0.806$, compared with the original value of 0.833. The optimal condition responds to the more rapid increase in pipe cost by decreasing the fraction of the total cost devoted to the pipe.

# 11-7 UNCONSTRAINED, MULTIVARIABLE OPTIMIZATION

The geometric-programming procedures for the one independent variable extend in a logical manner to multivariable optimizations. If applications continue to concentrate on problems of zero degree of difficulty, a two-variable problem, for example, will have an objective function containing three terms.

**Example 11-4** The pump and piping of Example 11-1 are actually part of a waste-treatment complex, as shown in Fig. 11-1. The system accomplishes the treatment by a combination of dilution and chemical action so that the effluent meets code requirements. The size of the reactor can decrease as the dilution increases. The cost of the reactor is $150/Q$, where $Q$ is the flow rate in cubic meters per second. The equation for the pumping cost with $Q$ broken out of the combined constant is $(220 \times 10^{15}Q^2)/D^5$. Use geometric programming to optimize the total system.

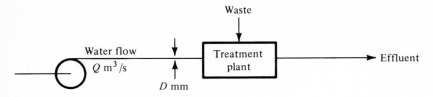

**Figure 11-1** Waste-treatment system in Example 11-1.

SOLUTION The total cost is the sum of the costs of the pipe, the pumping power, and the treatment plant

$$y = 160D + \frac{220 \times 10^{15}Q^2}{D^5} + \frac{150}{Q}$$

$$y^* = g^* = \left(\frac{160}{w_1}\right)^{w_1} \left(\frac{220 \times 10^{15}}{w_2}\right)^{w_2} \left(\frac{150}{w_3}\right)^{w_3}$$

provided that

$$w_1 + w_2 + w_3 = 1$$

to cancel

$D$: $$w_1 - 5w_2 = 0$$

$Q$: $$2w_2 - w_3 = 0$$

Solving gives

$$w_1 = \tfrac{5}{8} \qquad w_2 = \tfrac{1}{8} \qquad \text{and} \qquad w_3 = \tfrac{2}{8}$$

Then

$$\left(\frac{160}{5/8}\right)^{5/8} \left(\frac{220 \times 10^{15}}{1/8}\right)^{1/8} \left(\frac{150}{1/4}\right)^{1/4} = \$30,224$$

$$u_1^* = 160D^* = \tfrac{5}{8}(30,224) \qquad \text{so} \qquad D^* = 118 \text{ mm}$$

$$u_3^* = \frac{150}{Q^*} = \tfrac{2}{8}(30,224) \qquad \text{so} \qquad Q^* = 0.0198 \text{ m}^3/\text{s}$$

The core of the execution of Example 11-4 by geometric programming was the solution of *three* simultaneous *linear* equations for the $w$'s. Had the problem been solved by Lagrange multipliers, *two* simultaneous *nonlinear* equations (from $\nabla y = 0$) would have been solved. In general, the solution by geometric programming of a zero-degree-of-difficulty problem requires the solution of one more equation in a set of simultaneous equations than required by Lagrange multipliers, *but the equations are linear.* For this special class of problems, then, solution by geometric programming is likely to be much simpler than by calculus.

The mathematical substantiation for the procedure of solving the multi-variable problem, Example 11-4, has not yet been provided. It can be developed by following the pattern for the proof of the single-variable optimization presented in Sec. 11-5. The steps are as follows:

1. Propose a $g$ function in the form of Eq. (11-16).
2. Set the sum of the $w$'s equal to unity [Eq. (11-17)].
3. Maximize $\ln g$ subject to

$$\sum_{i=1}^{N} w_i = 1$$

to find that the optimal $w$'s equal the fractions that the respective $u$'s are of the total [Eq. (11-21)]. With those optimal values of $w$ the function $g$ equals the original function to be optimized.
4. Proceed to optimizing the function. The differentiation with respect to the $x$'s is now a *partial* differentiation, and the derivatives are equated to zero, following which the equations are multiplied through by the appropriate $x_i$.
5. The result will be a summation for each variable

$$\sum_{i=1}^{T} a_{tn} w_t = 0 \qquad (11\text{-}25)$$

where $t$ = term number in objective function
$T$ = total number of terms in objective function
$n$ = variable, ranging from 1 to total number of variables $N$

The $N$ equations designated by Eq. (11-25) are the conditions that result in the cancellation of all the $x$'s in the $g$-function formulation.


## 11-8 CONSTRAINED OPTIMIZATION WITH ZERO DEGREE OF DIFFICULTY

The final type of geometric-programming problem to be explored is one with an equality constraint. Only the zero-degree-of-difficulty case will be considered, and the total number of terms $T$ means the sum of those in the objective function and the constraint. Suppose that the objective function to be minimized is

$$y = u_1 + u_2 + u_3 \qquad (11\text{-}26)$$

subject to the constraint

$$u_4 + u_5 = 1 \qquad (11\text{-}27)$$

where the $u$'s are polynomials in terms of four independent variables, $x_1, x_2, x_3$, and $x_4$. The right side of the constraint equation must be unity, which poses no problem as long as one pure numerical term appears in the equation. If that number is not unity, the entire equation can be divided by the number to convert it into unity.

The objective function can be rewritten

$$y = g = \left(\frac{u_1}{w_1}\right)^{w_1} \left(\frac{u_2}{w_2}\right)^{w_2} \left(\frac{u_3}{w_3}\right)^{w_3} \tag{11-28}$$

provided that

$$w_1 + w_2 + w_3 = 1 \tag{11-29}$$

and

$$w_i = \frac{u_i}{u_1 + u_2 + u_3} \tag{11-30}$$

The constraint equation can also be rewritten as

$$u_4 + u_5 = 1 = \left(\frac{u_4}{w_4}\right)^{w_4} \left(\frac{u_5}{w_5}\right)^{w_5} \tag{11-31}$$

provided that

$$w_4 + w_5 = 1 \tag{11-32}$$

and

$$w_4 = \frac{u_4}{1} = u_4 \quad \text{and} \quad w_5 = \frac{u_5}{1} = u_5 \tag{11-33}$$

Equation (11-31) can be raised to the $M$th power, where $M$ is an arbitrary constant, and its value remains unity,

$$1 = \left(\frac{u_4}{w_4}\right)^{Mw_4} \left(\frac{u_5}{w_5}\right)^{Mw_5} \tag{11-34}$$

Next multiply Eq. (11-28) by Eq. (11-34)

$$y = g = \left(\frac{u_1}{w_1}\right)^{w_1} \left(\frac{u_2}{w_2}\right)^{w_2} \left(\frac{u_3}{w_3}\right)^{w_3} \left(\frac{u_4}{w_4}\right)^{Mw_4} \left(\frac{u_5}{w_5}\right)^{Mw_5} \tag{11-35}$$

Momentarily set aside the representations of Eqs. (11-28) to (11-35) and return to Eqs. (11-26) and (11-27) and solve by Lagrange multipliers

$$\nabla(u_1 + u_2 + u_3) - \lambda\left[\nabla(u_4 + u_5)\right] = 0 \tag{11-36}$$

$$u_4 + u_5 = 1 \tag{11-37}$$

The vector equation (11-36) represents four scalar equations, the terms of which are the partial derivatives with respect to $x_1$ to $x_4$. By taking advantage of the fact that all the $u$'s are polynomials each of the scalar equations of Eq. (11-36) can be multiplied by the variable with respect to which it has just been differentiated. The result is

$$a_{11}u_1^* + a_{21}u_2^* + a_{31}u_3^* - \lambda a_{41}u_4^* - \lambda a_{51}u_5^* = 0$$
$$\cdots\cdots\cdots\cdots\cdots\cdots\cdots\cdots\cdots\cdots\cdots\cdots\cdots\cdots$$
$$a_{14}u_1^* + a_{24}u_2^* + a_{34}u_3^* - \lambda a_{44}u_4^* - \lambda a_{54}u_5^* = 0 \tag{11-38}$$

The asterisk in Eqs. (11-38) indicates that these are optimal values of $u$. Next the results of Eqs. (11-38) can be merged with the representation of Eqs. (11-28) to (11-35). Dividing Eqs. (11-38) by $y^*$ permits replacement of $u_i^*/y^*$ by $w_i$.

Since the constant $M$ in Eq. (11-34) was arbitrary, let it equal $-\lambda/y^*$. The revised Eqs. (11-38) are then

$$a_{11}w_1 + a_{21}w_2 + a_{31}w_3 + Ma_{41}w_4 + Ma_{51}w_5 = 0$$
$$\cdots\cdots\cdots\cdots\cdots\cdots\cdots\cdots\cdots\cdots\cdots\cdots\cdots$$
$$a_{14}w_1 + a_{24}w_2 + a_{34}w_3 + Ma_{44}w_4 + Ma_{54}w_5 = 0 \qquad (11\text{-}39)$$

Along with the conditions

$$w_1 + w_2 + w_3 = 1 \qquad (11\text{-}29)$$

and, from Eq. (11-32),

$$Mw_4 + Mw_5 = M \qquad (11\text{-}40)$$

They provide six linear simultaneous equations. These six equations can be solved for the six unknowns $w_1$, $w_2$, $w_3$, $Mw_4$, $Mw_5$, and $M$.

It is especially significant that Eqs. (11-39) require the cancellation of all the $x$ terms in Eq. (11-35). This fact permits a convenient evaluation of the optimum value of $y$

$$y^* = \left(\frac{c_1}{w_1}\right)^{w_1} \left(\frac{c_2}{w_2}\right)^{w_2} \left(\frac{c_3}{w_3}\right)^{w_3} \left(\frac{c_4}{w_4}\right)^{Mw_4} \left(\frac{c_5}{w_5}\right)^{Mw_5}$$

**Example 11-5** A water pipeline extends 30 km across a desert from a desalination plant at the seacoast to a city. The pipeline, as shown schematically in Fig. 11-2, conveys 0.16 m$^3$/s of water. The first costs of the pipeline are

$$\text{Cost of each pump} = 2500 + 0.00032\Delta p^{1.2} \quad \text{dollars}$$

$$\text{Cost of 30 km of pipe} = 2{,}560{,}000D^{1.5} \quad \text{dollars}$$

where $\Delta p$ = pressure drop in each pipe section, Pa
$\quad D$ = diameter of pipe, m

Assume a friction factor of 0.02.

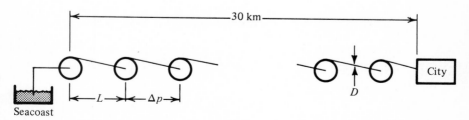

**Figure 11-2** Water pipeline in Example 11-5.

Use geometric programming on a constrained objective function to select the number of pumps and the pipe diameter that results in the minimum total first cost for the system.

SOLUTION If $n$ designates the number of pump-and-pipe sections, the total cost $y$ is

$$y = n(2500 + 0.00032\Delta p^{1.2}) + 2,560,000D^{1.5}$$

and

$$n = \frac{30,000 \text{ m}}{L}$$

where $L$ is the length of each pipe section in meters. The pressure drop in each pipe section is

$$\Delta p = f\frac{L}{D}\frac{V^2}{2}\rho = (0.02)\frac{L}{D}\left(\frac{0.16}{\pi D^2/4}\right)^2\frac{1}{2}(1000 \text{ kg/m}^3)$$

and so

$$\frac{\Delta p D^5}{L} = 0.4150 \qquad (11\text{-}41)$$

The statement of the problem is

Minimize
$$y = \frac{75,000,000}{L} + \frac{9.6\Delta p^{1.2}}{L} + 2,560,000D^{1.5} \qquad (11\text{-}42)$$

subject to
$$\frac{2.410\Delta p\, D^5}{L} = 1 \qquad (11\text{-}43)$$

$$y^* = \left(\frac{75,000,000}{w_1}\right)^{w_1}\left(\frac{9.6}{w_2}\right)^{w_2}\left(\frac{2,560,000}{w_3}\right)^{w_3}\left(\frac{2.41}{w_4}\right)^{Mw_4}$$

provided that

| | | |
|---|---|---|
| $L$: | $-w_1 - w_2$ | $- Mw_4 = 0$ |
| $\Delta p$: | $1.2w_2$ | $+ Mw_4 = 0$ |
| $D$: | | $1.5w_3 + 5Mw_4 = 0$ |
| | $w_1 + w_2 + w_3$ | $= 1$ |
| | | $Mw_4 = M$ |

so $w_1 = 0.0385, w_2 = 0.1923, w_3 = 0.7692, M = -0.2308$, and $Mw_4 = -0.2308$,

$$y^* = \left(\frac{75,000,000}{0.0385}\right)^{0.0385}\left(\frac{9.6}{0.1923}\right)^{0.1923}\left(\frac{2,560,000}{0.7692}\right)^{0.7692}\left(\frac{2.41}{1}\right)^{-0.2308}$$

$$y^* = \$410,150$$

$$u_1^* = (410,150)(0.0385) = \frac{75,000,000}{L}$$

and so
$$L^* = 4750 \text{ m}$$

$$u_3^* = (410{,}150)(0.762) = 2{,}560{,}000D^{1.5}$$

and so
$$D^* = 0.246 \text{ m}$$

$$\Delta p^* = \frac{L^*}{2.410D^{*5}} = 2{,}188{,}000 \text{ Pa} = 2188 \text{ kPa}$$

Since the number of pump-and-pipe sections is 30,000/4750 = 6.3, six pumps could be used, which revises $L$ to 5000 m, and this value of $L$ can be substituted into Eqs. (11-42) and (11-43) for a reoptimization of $\Delta p$ and $D$.

## 11-9  SENSITIVITY COEFFICIENT

In Sec. 8-10 the Lagrange multiplier $\lambda$ emerged as the sensitivity coefficient, which was $dy^*/dA$, where $A$ is the numerical term on the right side of the constraint equation. In the development of the procedure for constrained optimization using geometric programming the expression $-\lambda/y^*$ was replaced by the term $M$ in arriving at Eqs. (11-39). When the optimization is complete and $y^*$ and $M$ are known, the sensitivity coefficient can be determined quickly. In Example 11-5 $M = -0.2308$ and $y^* = 410{,}150$, so the sensitivity coefficient $\lambda = -My^* = 94{,}660$. This value of the sensitivity coefficient is applicable to the constraint equation in the form of Eq. (11-43). The value of the sensitivity coefficient for the original form of the constraint in Eq. (11-41) is 94,660/0.4150 = 228,100.

The physical interpretation of the sensitivity coefficient in Example 11-5 is that it indicates the rate of change of the total first cost with respect to the change in the constant 0.4150. If, for example, the water flow rate were changed from 0.160 to 0.161 m$^3$/s, the constant would change from 0.4150 to (0.4150) $(0.161/0.160)^2 = 0.4202$. The increase in cost of the system is (228,100) (0.4202 − 0.4150) = $1187.

## 11-10  HIGHER DEGREES OF DIFFICULTY AND EXTENSIONS OF GEOMETRIC PROGRAMMING

The only optimization problems to which geometric programming was applied in this chapter were those of zero degree of difficulty. For such problems geometric programming is probably the simplest means of solution. Geometric programming also provides some physical insight into the solution not usually offered by other methods of optimization.

Geometric programming is not limited to zero-degree-of-difficulty problems but can accommodate higher degrees of difficulty. The need to solve a set of nonlinear simultaneous equations arises in problems of nonzero degree of difficulty. Much of the effort by operations-research workers who are concentrating on geometric programming is devoted to developing efficient and reliable computer programs that include the solution of these nonlinear equations. These computer programs can also handle inequality constraints, as well as the equality constraints presented in this chapter.

Geometric programming is a useful tool to carry in the optimization kit for those special situations to which it is particularly adaptable.

## PROBLEMS

Solve the following problems by geometric programming.

**11-1** The thickness of the insulation of a hot-water tank is to be selected so that the total cost of the insulation and standby heating for the 10-year life of the facility will be minimum.

*Data*

Average water temperature, $60°C$
Average ambient temperature, $24°C$
Conductivity of insulation, $0.035$ W/(m · K)
Cost of heat energy, $4 per gigajoule
Cost of insulation, where $x$ = insulation thickness, mm, $0.5x^{0.8}$ dollars per square meter

The operation is continuous. Assume that the only resistance to heat transfer is the insulation.

(a) What is the minimum total cost of insulation plus standby heat loss per square meter of heat-transfer area for 10 years, neglecting interest charges?

(b) What is the optimum insulation thickness?

Ans.: (b) 99.3 mm.

**11-2** A hydraulic power system must provide 300 W of power, where the power is the product of the volume flow rate $Q$ m³/s and the pressure buildup $\Delta p$ Pa. The cost of the hydraulic pump is a function of both the flow rate and pressure buildup:

$$\text{Cost} = 1200Q^{0.4}\sqrt{10 + (\Delta p \times 10^{-4})} \quad \text{dollars}$$

Convert to a single-variable unconstrained problem and use geometric programming to determine the minimum cost of the pump and the optimum values of $Q$ and $\Delta p$.

Ans.: $\Delta p^* = 400$ kPa.

**11-3** Schlichting[5] presents the following equation for the velocity in a free-stream jet that issues into a large space, as illustrated in Fig. 11-3:

$$u = \frac{7.41u_o\sqrt{A}}{x[1 + 57.5(r/x)^2]^2}$$

where $u$ = velocity in the $x$ direction, m/s
$r$ = radial distance measured from the centerline, m
$u_o$ = outlet velocity, m/s
$A$ = outlet area, m²
$x$ = distance in $x$ direction measured from wall, m

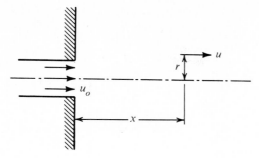

**Figure 11-3** Free-stream jet in Prob. 11-3.

The position of maximum $u$ at a radial distance of 0.5 m from the centerline is to be determined in a certain jet with given values of $u_o$ and $A$. Specifically, use geometric programming to find the value of $x$ at which $u$ is maximum when $r = 0.5$ m. *Hint*: The problem can be structured as one of zero degree of difficulty.

Ans.: 6.57 m.

**11-4** A hot-water boiler consists of a combustion chamber and a heat exchanger arranged as shown in Fig. 11-4. Fuel at a flow rate of 0.0025 kg/s with a heating value of 42,000 kJ/kg burns in the combustion chamber. Thereafter the flue gases are cooled to 150°C in the process of heating the water. The combustion efficiency increases with an increase in the rate of airflow according to the equation

$$\text{Efficiency, \%} = 100 - \frac{0.023}{(m_a + 0.0025)^2}$$

The specific heat of the gas mixture is 1.05 kJ/(kg · K). Determine the value of $m_a$ that results in the maximum rate of heat transfer to the water. *Suggestion*: Use $m_a + 0.0025$ as the variable to be optimized.

Ans.: 0.0695 kg/s.

**11-5** In a three-stage compression system air enters the first stage at a pressure of 100 kPa and a temperature of 20°C. The exit pressure from the third stage is 6400 kPa. Between stages the air is passed through an intercooler that brings the temperature back to 20°C. The expression for the work of compression per unit mass in an ideal process is

$$\frac{k}{k-1}RT_1 \left[1 - \left(\frac{p_2}{p_1}\right)^{(k-1)/k}\right]$$

where subscript 1 refers to the entering conditions and subscript 2 to the leaving conditions.

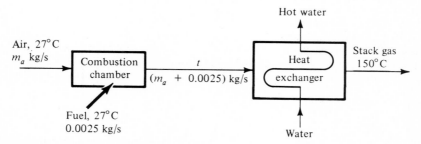

**Figure 11-4** Hot-water boiler in Prob. 11-4.

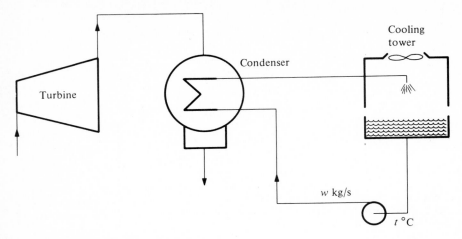

**Figure 11-5** Heat-rejection system in Prob. 11-6.

(a) With the intermediate pressures chosen so that the total work of compression is a minimum, what is the minimum work required to compress 1 kg of air, assuming that the compressions are reversible and adiabatic?

(b) What are the intermediate pressures that result in this minimum work of compression?

**Ans.:** 429 kJ; 400 and 1600 kPa.

**11-6** The heat-rejection system for a condenser of a steam power plant (Fig. 11-5) is to be designed for minimum first plus pumping cost. The heat-rejection rate from the condenser is 14 MW. The following costs in dollars must be included:

First cost of cooling tower, $800A^{0.6}$, where $A$ = area, $m^2$
Lifetime pumping cost, $0.0005w^3$, where $w$ = flow rate of water, kg/s
Lifetime penalty in power production due to elevation of temperature of cooling water, $270t$, where $t$ = temperature of water entering the condenser, °C
The rate of heat transfer from the cooling tower can be represented adequately by the expression $q$, W = $3.7(w^{1.2})tA$.

(a) Set up an unconstrained objective function in terms of the variables $A$ and $w$.
(b) Determine the minimum lifetime cost.
(c) Calculate the optimal values of $A$ and $w$.

**Ans.:** (c) $w$ = 202.6 kg/s.

**11-7** The total cost of a rectangular building shell and the land it occupies is to be minimized for a building that must have a volume of 14,000 $m^3$. The following costs per square meter apply: land, $90; roof, $14; floor, $8; and walls, $11. Set up the problem as one of constrained optimization and determine the minimum cost and the optimal dimensions of the building.

**Ans.:** Height = 71.4 m.

**11-8** Newly harvested grain often has a high moisture content and must be dried to prevent spoilage. This drying can be achieved by warming ambient air and blowing it through a bed of the grain. The seasonal operating cost in dollars per square meter of grain bed for such a dryer consists of the cost of heating the air

$$\text{Heating cost} = 0.002Q \, \Delta t$$

and the blower operating cost

$$\text{Blower cost} = 2.6 \times 10^{-9}Q^3$$

where $Q$ = air quantity delivered through the bed during season, $\text{m}^3/\text{m}^2$ bed area
$\Delta t$ = rise in air temperature through heater, °C

The values of $Q$ and $\Delta t$ also influence the time required for adequate drying of the grain according to the equation

$$\text{Drying time} = \frac{80 \times 10^6}{Q^2 \Delta t} \quad \text{days}$$

Using the geometric-programming method of constrained optimization, compute the minimum operating cost and the optimum value of $Q$ and $\Delta t$ that will achieve adequate drying in 60 days.

Ans.: $\Delta t = 2.28°\text{C}$.

# REFERENCES

1. R. J. Duffin, E. L. Peterson, and C. M. Zener, *Geometric Programming*, Wiley, New York, 1967.
2. W. F. Stoecker, *Design of Thermal Systems*, 1st ed., McGraw-Hill, New York, 1971.
3. C. Zener, *Engineering Design by Geometric Programming*, Wiley-Interscience, New York, 1971.
4. C. Beightler and D. T. Phillips, *Applied Geometric Programming*, Wiley, New York, 1976.
5. H. Schlichting, *Boundary Layer Theory*, 5th ed., McGraw-Hill, New York, 1958.

# TWELVE

## LINEAR PROGRAMMING

### 12-1 THE ORIGINS OF LINEAR PROGRAMMING

Linear programming is an optimization method applicable where both the objective function and the constraints can be expressed as linear combinations of the variables. The constraint equations may be equalities or inequalities. Linear programming first appeared in Europe in the 1930s, when economists and mathematicians began working on economic models. During World War II the United States Air Force sought more effective procedures for allocating resources and turned to linear programming. In 1947 a member of the group working on the Air Force problem, George Dantzig, reported the *simplex method* for linear programming, which was a significant step in bringing linear programming into wider use.

Economists and industrial engineers have applied linear programming more in their fields of work than most other technical groups. Decisions about time allocations of machines to various products in a manufacturing plant, for example, lend themselves neatly to linear programming. In thermal systems linear programming has become an important tool in the petroleum industry and is now being applied in other thermal industries. Most large oil companies use linear-programming models to determine the quantities of the various products that will result in optimum profit for the entire operation. Within the refinery itself, linear programming helps determine where the bottlenecks to production exist and how much the total output of the plant could be increased, for instance, by enlarging a heat exchanger or by increasing a certain rate of flow.

Entire books are devoted to linear programming, but it is possible in one chapter to explain the physical situation that results in a linear-programming structure and to solve practical problems with the technique. The emphasis will be on obtaining a geometric feel for the linear-programming situation, and the simplex algorithm for solving both maximization and minimization problems in linear programming will be applied.

## 12-2 SOME EXAMPLES OF LINEAR PROGRAMMING

Some classic uses of linear programming are to solve (1) the blending problem, (2) machine allocation, (3) inventory and production planning, and (4) the transportation problem.

The oil-company application mentioned in Sec. 12-1 is typical of the blending application. The oil company has a choice of buying crude from several different sources with different compositions and at differing prices. It has a choice of manufacturing various quantities of aviation fuel, automobile gasoline, diesel fuel, and oil for heating. The combinations of these products are restricted by material balances based on the incoming crude and by the capacity of such components in the refinery as the cracking unit. A mix of purchased crude and manufactured products is sought that gives maximum profit.

The machine-allocation problem occurs where a manufacturing plant has a choice of making several different products, each of which requires varying machine times of different machines such as lathes, screw machines, and grinders. The machine time of some or all of these different machines is limited. The goal of the analysis is to determine the production quantities of each product that result in maximum profit.

The sales of some manufacturing firms fluctuate, often according to a seasonal pattern. The company can build up an inventory of manufactured products to carry it through the period of peak sales, but carrying an inventory costs money. Or it can pay overtime rates in order to step up its production during the period of peak sales, which also entails an additional expense. Finally, the company can simply plan on losing some sales because it does not meet the sales demand at the time that it exists, thus losing a potential profit. Linear programming can incorporate the various cost and loss factors and arrive at the most profitable production plan.

The fourth application, the transportation problem, occurs when an organization has several production plants distributed throughout a geographical area and a number of differently distributed warehouses. Each plant has a certain production capability, and each warehouse has a certain requirement. Any one warehouse may receive the production from one or more plants. The object is to determine how much of each plant's production should be shipped to each warehouse in order to minimize the total manufacturing and transportation cost.

Simple linear-programming problems can be done in a hit-or-miss fashion, but those with three or more variables require systematic procedures. Even when using methodical techniques, the magnitude of a problem that can be solved by hand is limited. Large problems (with several thousand variables) require computer programs, which are currently available as library routines.

## 12-3 MATHEMATICAL STATEMENT OF THE LINEAR-PROGRAMMING PROBLEM

The form of the statement is typical of the optimization problem in that it consists of an objective function and constraints. The objective function which is to be minimized (or maximized) is

$$y = c_1 x_1 + c_2 x_2 + \cdots + c_n x_n \tag{12-1}$$

and the constraints are

$$\phi_1 = a_{11} x_1 + a_{12} x_2 + \cdots + a_{1n} x_n \geqslant r_1$$
$$\cdots\cdots\cdots\cdots\cdots\cdots\cdots\cdots\cdots\cdots\cdots\cdots\cdots \tag{12-2}$$
$$\phi_m = a_{m1} x_1 + a_{m2} x_2 + \cdots + a_{mn} x_n \geqslant r_m$$

Furthermore, if the $x$'s represent physical quantities, they are likely to be non-negative, so that $x_1, \ldots, x_n \geqslant 0$. The $c$ values and $a$ values are all constants, which make both the objective function and the constraints linear; hence the name linear programming. The values of $c$ and $a$ may be positive, negative, or zero. The inequalities in the constraints can be in either direction and can even be strict equalities.

At first glance, this problem might seem readily soluble by Lagrange multipliers, but we recall that the method of Lagrange multipliers is applicable where *equality* constraints exist. Furthermore, Lagrange multipliers apply where $n > m$, but in linear programming $n$ can be greater than, equal to, or less than $m$. The significance of $n < m$ will be discussed in Sec. 12-13.

## 12-4 DEVELOPING THE MATHEMATICAL STATEMENT

The translation of the physical conditions into a mathematical statement of linear-programming form will be illustrated by an example.

**Example 12-1** A simple power plant consists of an extraction turbine that drives a generator, as shown in Fig. 12-1. The turbine receives 3.2 kg/s of steam, and the plant can sell either electricity or extraction steam for processing purposes. The revenue rates are

Electricity, $0.03 per kilowatthour
Low-pressure steam, $1.10 per megagram
High-pressure steam, $1.65 per megagram

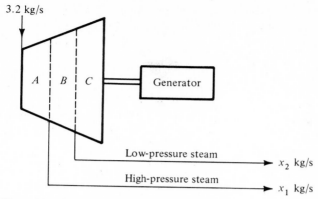

**Figure 12-1** Power plant in Example 12-1.

The generation rate of electric power depends upon the flow rate of steam passing through each of the sections $A$, $B$, and $C$; these flow rates are $w_A$, $w_B$, and $w_C$, respectively. The relationships are

$$P_A, \text{kW} = 48w_A$$
$$P_B, \text{kW} = 56w_B$$
$$P_C, \text{kW} = 80w_C$$

where the $w$'s are in kilograms per second. The plant can sell as much electricity as it generates, but there are other restrictions.

To prevent overheating the low-pressure section of the turbine, no less than 0.6 kg/s must always flow through section $C$. Furthermore, to prevent unequal loading on the shaft, the permissible combination of extraction rates is such that if $x_1 = 0$, then $x_2 \leqslant 1.8$ kg/s, and for each kilogram of $x_1$ extracted 0.25 kg less can be extracted of $x_2$.

The customer of the process steam is primarily interested in total energy and will purchase no more than

$$4x_1 + 3x_2 \leqslant 9.6$$

Develop the objective function for the total revenue from the plant and also the constraint equations.

SOLUTION The revenue per hour is the sum of the revenues from selling the steam and the electricity.

$$\text{Revenue} = \frac{1.65}{1000}(3600x_1) + \frac{1.10}{1000}(3600x_2) + 0.03(48w_A + 56w_B + 80w_C)$$

Since $w_A = 3.2$ kg/s and from mass balances $w_B = 3.2 - x_1$ and $w_C = 3.2 - x_1 - x_2$,

$$\text{Revenue} = 17.66 + 1.86x_1 + 1.56x_2 \tag{12-3}$$

Because the constant has no effect on the state point at which the optimum occurs, the objective function to be maximized is

$$y = 1.86x_1 + 1.56x_2 \tag{12-4}$$

The three constraints are

$$x_1 + x_2 \leqslant 2.6 \tag{12-5}$$

$$x_1 + 4x_2 \leqslant 7.2 \tag{12-6}$$

$$4x_1 + 3x_2 \leqslant 9.6 \tag{12-7}$$

## 12-5 GEOMETRIC VISUALIZATION OF THE LINEAR-PROGRAMMING PROBLEM

Since it involves only the two variables $x_1$ and $x_2$, Example 12-1 can be illustrated geometrically as in Fig. 12-2. The constraint of Eq. (12-5), for example, states that only the region on and to the left of the line $x_1 + x_2 = 2.6$ is permitted. Placing the other two constraints in Fig. 12-2 further restricts the permitted region to *ABDFG*.

Next the lines of constant revenue $y$ are plotted on Fig. 12-2. Inspection shows that the greatest profit can be achieved by moving to point $D$, where $x_1 = 1.8$ and $x_2 = 0.8$. An important generalization is that the *optimum solution lies at a corner*. A special case of this generalization is where the line of constant profit is parallel to a constraint line, in which case any point on the constraint line between the corners is equally favorable.

If the objective function depends upon three variables, a three-dimensional

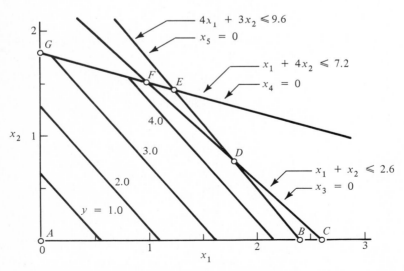

**Figure 12-2** Constraints and lines of constant profit in Example 12-1.

graph is required, and then the constraint equations are represented by planes. The corner where the optimum occurs is formed by the intersection of three planes.

## 12-6  INTRODUCTION OF SLACK VARIABLES

The constraint equations (12-5) to (12-7) are inequalities, but they can be converted into equalities by the introduction of another variable in each equation

$$x_1 + x_2 + x_3 \qquad\qquad = 2.6 \qquad\qquad (12\text{-}8)$$

$$x_1 + 4x_2 \qquad + x_4 \qquad = 7.2 \qquad\qquad (12\text{-}9)$$

$$4x_1 + 3x_2 \qquad\qquad + x_5 = 9.6 \qquad\qquad (12\text{-}10)$$

This substitution is valid provided that $x_3 \geqslant 0$, $x_4 \geqslant 0$, and $x_5 \geqslant 0$. These new variables are called *slack variables*.

Reference to Fig. 12-2 permits a geometric interpretation of the slack variables $x_3$, $x_4$, and $x_5$. Any point on the graph defines specific values of $x_3$, $x_4$, and $x_5$. Along the $x_1 + x_2 = 2.6$ line, for example, $x_3 = 0$. The value of $x_3$ in the region to the right of the line is less than zero and is thus prohibited, while on the line and to the left of it $x_3 \geqslant 0$, and thus this region is permitted.

## 12-7  PREPARATION FOR THE SIMPLEX ALGORITHM

The simplex algorithm has a mathematical basis, but the mechanics will be presented before the theory. No rigorous proof will be given, but a geometric explanation will be provided to give insight into the functions the simplex algorithm is performing.

The first step in preparing for the simplex algorithm is to write the equations in table or tableau form. The reason is that equations will be written many times; instead of repeating $x_1$, for example, in all the equations $x_1$ will be used as a column heading, as in Table 12-1. The double line is interpreted as the equality sign and only the coefficients of the $x$ terms appear in the boxes.

Table 12-1  Constraint equations in tableau form

| $x_1$ | $x_2$ | $x_3$ | $x_4$ | $x_5$ | |
|---|---|---|---|---|---|
| 1 | 1 | 1 | | | 2.6 |
| 1 | 4 | | 1 | | 7.2 |
| 4 | 3 | | | 1 | 9.6 |
| | | | | | |

## 12-8 INCLUDING THE OBJECTIVE FUNCTION IN THE TABLEAU

The bottom line of Table 12-1 was left blank anticipating including the objective function, Eq. (12-4). The form of the equation is revised, however, before inserting the numbers so that all the $x$ terms are moved to the left side of the equation

$$y - 1.86x_1 - 1.56x_2 = 0 \qquad (12\text{-}11)$$

Insertion of these terms into the tableau yields the results shown in Table 12-2. The coefficients of the $x$ terms in the bottom line for the objective function are called *difference coefficients*.

## 12-9 STARTING AT THE ORIGIN

The progression in linear-programming solutions is to move from one corner to the next corner, achieving an improvement in the objective function with each move. When no further improvement is possible, the optimum has been reached. The starting point is always the origin, namely, the position where the physical variables are zero. In Example 12-1 the starting point is where $x_1 = 0$ and $x_2 = 0$. One technique of indicating these values is to note them in the column heading; this converts the tableau into the form of Table 12-3, which is now the complete tableau 1.

**Table 12-2 Constraint equations and objective function in tableau form**

| $x_1$ | $x_2$ | $x_3$ | $x_4$ | $x_5$ | |
|-------|-------|-------|-------|-------|-----|
| 1 | 1 | 1 | | | 2.6 |
| 1 | 4 | | 1 | | 7.2 |
| 4 | 3 | | | 1 | 9.6 |
| −1.86 | −1.56 | | | | 0 |

**Table 12-3 Tableau 1 of Example 12-1**

| $x_1 = 0$ | $x_2 = 0$ | $x_3$ | $x_4$ | $x_5$ | |
|-----------|-----------|-------|-------|-------|-----|
| 1 | 1 | 1 | | | 2.6 |
| 1 | 4 | | 1 | | 7.2 |
| 4 | 3 | | | 1 | 9.6 |
| −1.86 | −1.56 | | | | 0 |

A property of all tableaux throughout the linear-programming procedure is that the current values of all the $x$'s and the objective function can be read immediately from the tableau. In Table 12-3, for example, $x_1$ and $x_2$ are zero; since the first line corresponds to Eq. (12-8), the value of $x_3$ is 2.6. Similarly, $x_4 = 7.2$ and $x_5 = 9.6$. The bottom line is the objective function in the form of Eq. (12-11), so $y = 0$. The number in the lower right corner of each tableau is the current value of the objective function.

## 12-10  THE SIMPLEX ALGORITHM

The simplex algorithm is a procedure whereby the successive tableaux can be developed from the first tableau. The steps are as follows:

1. Decide which of the variables that currently are zero should be programmed next. In a maximization problem the variable with the largest negative difference coefficient is chosen; in minimization the variable with the largest positive difference coefficient is chosen.
2. Determine which is the controlling constraint by selecting the constraint with the most restrictive (the smallest) quotient of the numerical term on the right side of the equality divided by the coefficient of the variable being programmed.
3. Transfer the controlling constraint to the new tableau by dividing all terms by the coefficient of the variable being programmed.
4. For all other boxes in the new tableau (including noncontrolling constraints and difference coefficients) use the following procedure:
   a. Select a box in the new tableau. Call the value in the same box of the old tableau $v$.
   b. Move sideways in the old tableau to the coefficient of the variable being programmed. Call this value $w$.
   c. In the new tableau move from the box being calculated up or down to the row which contains the previous controlling equation. Call the value in that box $z$.
   d. The value of the box in the new tableau is $v - wz$.

## 12-11  SOLUTION OF EXAMPLE 12-1

The simplex algorithm described in Sec. 12-10 will now be used to solve Example 12-1. The first tableau is complete as shown in Table 12-3 and is reproduced in Table 12-4 for transformation to the second tableau.

***Step 1*** The first step is to decide which of the variables currently noted in the column heading as zero ($x_1$ or $x_2$) should be programmed first. The largest negative difference coefficient is $-1.86$ in the $x_1$ column, which indicates that $x_1$

**Table 12-4 Tableau 1 of Example 12-1 with indication of variable being programmed and controlling equation**

|  | $x_1 = 0$ | $x_2 = 0$ | $x_3$ | $x_4$ | $x_5$ |  |
|---|---|---|---|---|---|---|
| 2.6/1 = 2.6 | 1 | 1 | 1 |  |  | 2.6 |
| 7.2/1 = 7.2 | 1 | 4 |  | 1 |  | 7.2 |
| ⟹9.6/4 = 2.4 | 4 | 3 |  |  | 1 | 9.6 |
|  | −1.86 | −1.56 |  |  |  | 0 |

should be programmed (increased from its zero value). The vertical arrow indicates that $x_1$ is being programmed.

**Step 2** The next step is to determine to what extent $x_1$ can be increased. The numerical terms to the right of the equality sign are divided by the coefficients of the variable being programmed in the same line, 2.6/1, 7.2/1, and 9.6/4. The smallest of the resulting quotients is 2.4, as shown in the left column of Table 12-4, so the third constraint is the controlling one.

In Fig. 12-2 the operation in moving from tableau 1 to tableau 2 is that of moving along the $x_1$ axis from $A$ to $B$. The quotients in the left column denote values of $x_1$ that the respective constraints permit. The first constraint would permit $x_1$ to increase to 2.6, but the most restrictive is the third constraint, which permits $x_1$ to increase to 2.4.

The construction of tableau 2 in Table 12-5 can now begin with the designation of the column headings. The variable $x_1$ was zero but is no longer, and $x_2$ remains zero. The variable $x_5$, which was nonzero, now becomes zero, as indicated both by the geometry in Fig. 12-2 and by the third constraint in tableau 1. If that constraint is thought of as an equation, the only variables taking part in

**Table 12-5 Tableau 2 of Example 12-1**

|  | $x_1$ | $x_2 = 0$ | $x_3$ | $x_4$ | $x_5 = 0$ |  |
|---|---|---|---|---|---|---|
| ⟹0.8 | 1 − (1)(1) = 0 | 1 − (1)(0.75) = 0.25 | 1 − (1)(0) = 1 | 0 − (1)(0) = 0 | 0 − (1)(0.25) = −0.25 | 2.6 − (1)(2.4) = 0.20 |
| 1.48 | 1 − (1)(1) = 0 | 4 − (1)(0.75) = 3.25 | 0 − (1)(0) = 0 | 1 − (1)(0) = 1 | 0 − (1)(0.25) = −0.25 | 7.2 − (1)(2.4) = 4.8 |
| 3.2 | 1 | 0.75 | 0 | 0 | 0.25 | 2.4 |
|  | −1.86 − (−1.86)(1) = 0 | −1.56 − (−1.86)(0.75) = −0.165 | 0 − (−1.86)(0) = 0 | 0 − (−1.86)(0) = 0 | 0 − (−1.86)(0.25) = 0.465 | 0 − (−1.86)(2.4) = 4.464 |

the action of the move are $x_1$ and $x_5$. All the other variables are either zero or have zero coefficients. The variable $x_1$ increases until $x_5$ has dropped to zero.

**Step 3** The first numbers inserted in the boxes of tableau 2 come from the controlling equation of tableau 1. Dividing all the coefficients of the third constraint by 4, which is the coefficient of the variable being programmed, the numbers in the boxes of tableau 2 for the third constraint become 1, 0.75, 0, 0, 0.25, and 2.4.

**Step 4** For all the other boxes in tableau 2 the $v - wz$ routine is followed. The individual calculations are shown in the boxes of tableau 2.

Tableau 2 is now complete. The information that can be read off it is the values of all the $x$'s and the value of the objective function. Two of the $x$'s, namely $x_2$ and $x_5$, are zero, as shown by the column heading. For each of the other $x$'s a constraint equation will provide its value. In the first constraint all the variables are either zero or have a zero coefficient, except for $x_3$. So the equation has reduced to $x_3 = 0.20$. From the other two constraints $x_4 = 4.8$ and $x_1 = 2.4$. The box in the bottom right corner indicates the current value of the objective function to be 4.464.

The entire simplex algorithm is repeated to transform tableau 2 into tableau 3. The largest negative difference coefficient, in fact the only negative one, is $-0.165$, under $x_2$, so $x_2$ is programmed next. This time the first constraint is most restrictive; it therefore becomes the controlling equation, and as $x_2$ increases to its limit, $x_3$ drops to zero, as shown by the column heading in tableau 3 (Table 12-6). The coefficients in the first constraint of tableau 2 are divided by 0.25 and the terms transferred to tableau 3. The remaining boxes are computed according to step 4 in Sec. 12-10. The complete tableau 3 shows $x_1 = 1.8$, $x_2 = 0.8$, $x_3 = 0$, $x_4 = 2.2$, $x_5 = 0$, and $y = 4.596$.

The search for which variable to program next discloses that there are no negative difference coefficients, and this condition indicates that no further improvement is possible and the optimum has been reached. The transformation from tableau 2 to tableau 3 was a move on Fig. 12-2 from point $B$ to point $D$. At the solution two of the slack variables, $x_3$ and $x_5$, are zero, but $x_4 = 2.2$. These values indicate that the second constraint has no influence on the optimum, although it might not have been possible to realize this fact in advance.

**Table 12-6 Tableau 3 of Example 12-1**

| $x_1$ | $x_2$ | $x_3 = 0$ | $x_4$ | $x_5 = 0$ | |
|-------|-------|-----------|-------|-----------|------|
| 0 | 1 | 4 | 0 | $-1$ | 0.8 |
| 0 | 0 | $-13$ | 1 | 3 | 2.2 |
| 1 | 0 | $-3$ | 0 | 1 | 1.8 |
| 0 | 0 | 0.66 | 0 | 0.3 | 4.596 |

## 12-12 ANOTHER GEOMETRIC INTERPRETATION OF THE TABLEAU TRANSFORMATION

Figure 12-2 showed one geometric interpretation of the linear-programming problem and the progressive moves from one corner to another. Another geometric interpretation can be achieved by changing the coordinates with each transformation so that the current point is always at the origin. Since the first tableau starts at the origin of the physical variables, the coordinates $x_1$ and $x_2$ in Fig. 12-2 represent the first tableau. In moving from tableau 1 to tableau 2 the coordinate $x_1$ is replaced by $x_5$. The procedure in making this replacement is to solve for $x_1$ in the third constraint

$$x_1 = 2.4 - 0.75x_2 - 0.25x_5 \qquad (12\text{-}12)$$

and substitute this expression for $x_1$ into the first two constraints and the objective function to obtain

Constraint 1: $\qquad\qquad 0.25x_2 + x_3 - 0.25x_5 = 0.2 \qquad\qquad (12\text{-}13)$

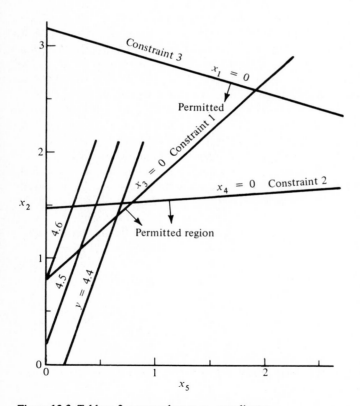

**Figure 12-3** Tableau 2 expressed on $x_5 x_2$ coordinates.

Constraint 2:  $\qquad 3.25x_2 + x_4 - 0.25x_5 = 4.8$  (12-14)

and  $\qquad y - 0.165x_2 + 0.465x_5 = 4.464$  (12-15)

The three constraints, Eqs. (12-12) to (12-14), are shown on the $x_2 x_5$ coordinates of Fig. 12-3, as well as a few lines of constant objective function from Eq. (12-15). An examination of Fig. 12-3 indicates that to improve (increase) the objective function $x_2$ should be increased until a constraint limits the advance. The restriction is imposed by constraint 1, which occurs when $x_3 = 0$, so the coordinate $x_2$ will now be replaced by $x_3$. That replacement will be performed by solving for $x_2$ in Eq. (12-13) and substituting into Eqs. (12-12), (12-14), and (12-15). The resulting set of equations is

Constraint 1:  $\qquad x_2 + 4x_3 - x_5 = 0.8$  (12-16)

Constraint 2:  $\qquad -13x_3 + x_4 + 3x_5 = 2.2$  (12-17)

Constraint 3:  $\qquad x_1 - 3x_3 + x_5 = 1.8$  (12-18)

Objective function:  $\quad y + 0.66x_3 + 0.3x_5 = 4.596$  (12-19)

A graph on $x_5 x_3$ coordinates of the constraints and constant values of the objective function is shown in Fig. 12-4. The current point is once again located at the origin, and any attempt to increase the objective function is doomed because an increase in either $x_3$ or $x_5$ decreases the value. No further improvement is possible, and the optimum has been reached.

The particularly significant point about the transformations where the coordinates were progressively replaced in Figs. 12-3 and 12-4 is the equations on which the graphs are based. If Eqs. (12-13) and (12-14) to (12-15) are placed in a tableau form with the $x$'s as the column headings, Table 12-7 is the result. A comparison of Table 12-7 with tableau 2 (Table 12-5) shows the two tables to be identical. Equations (12-16) to (12-19) could also be tabulated to show that these equations reproduce tableau 3. The conclusion is that the linear-programming process can be visualized as a progressive transformation of co-

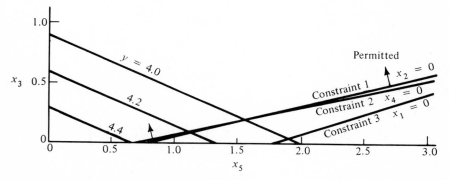

**Figure 12-4**  Tableau 3 expressed on $x_5 x_3$ coordinates.

**Table 12-7 Equations (12-12) to (12-15)**

| $x_1$ | $x_2$ | $x_3$ | $x_4$ | $x_5$ | |
|---|---|---|---|---|---|
| 0 | 0.25 | 1 | | −0.25 | 0.2 |
| 0 | 3.25 | 0 | 1 | −0.25 | 4.8 |
| 1 | 0.75 | | | 0.25 | 2.4 |
| 0 | −0.165 | | | 0.465 | 4.464 |

ordinates where in each transformation one of the coordinates is replaced by a variable indicating the limiting constraint.

## 12-13 NUMBER OF VARIABLES AND NUMBER OF CONSTRAINTS

The relationship of the number of physical variables and the number of constraints gives an indication of the number of variables that are zero in the solution. Let the number of physical variables be denoted by $n$ and the number of constraints (and therefore the number of slack variables) be denoted by $m$. There will always be $n$ variables (physical plus slack) equal to zero at the optimum, or for that matter at any corner. When $m > n$, as Fig. 12-5a shows, at least $m - n$ constraints play no role in the solution. In Fig. 12-5b where $m < n$, at least $n - m$ physical variables are zero.

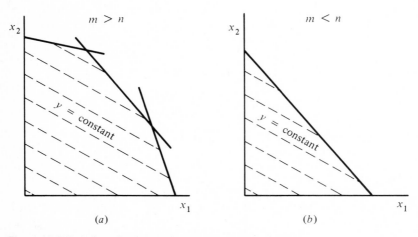

**Figure 12-5** Relation of number of physical and slack variables.

## 12-14 MINIMIZATION WITH GREATER-THAN CONSTRAINTS

Solution of the maximization problem with less-than constraints consisted of moving from one corner in the feasible region to whichever adjacent corner showed the most improvement in the objective function. Since linear programming always starts at the physical origin, the origin is in the feasible region with less-than constraints. In the minimization problem with greater-than constraints, locating the first feasible point may be difficult. Admittedly, in simple problems involving a small number of variables, combinations of variables could be set to zero in the constraint equations and the other variables solved until a combination is found that violates no constraints. In large problems this method is prohibitive and a more systematic procedure must be employed. The introduction of *artificial variables* facilitates this procedure.

## 12-15 ARTIFICIAL VARIABLES

Suppose that an inequality constraint with a greater-than sense is to be converted into an equality but also must permit the physical variables to take on zero values. The inequality

$$3x_1 + 4x_2 \geqslant 12 \tag{12-20}$$

can be converted into an equality by introducing a slack variable $x_3$,

$$3x_1 + 4x_2 - x_3 = 12 \tag{12-21}$$

The slack variable takes on a negative sign, so that the constraint is satisfied when $x_3 \geqslant 0$.

The next requirement is the ability to set $x_1$ and $x_2$ equal to zero. Equation (12-21) will not permit zero values of $x_1$ and $x_2$, and so another variable, the artificial variable, is inserted

$$3x_1 + 4x_2 - x_3 + x_4 = 12 \tag{12-22}$$

The geometric interpretation taken on by the slack and artificial variables with respect to the constraint is shown in Fig. 12-6. Along the constraint $3x_1 + 4x_2 = 12$, $x_3 = 0$, and $x_4 = 0$. When moving to the right of the constraint into the feasible region, $x_3$ takes on positive values and $x_4$ remains zero. When moving to the left of the constraint into the infeasible region, $x_4$ takes on positive values and $x_3$ remains zero. The result of the introduction of both the slack and artificial variable is that the position may be located anywhere on the graphs but all the variables are still abiding by the requirement that $x_i \geqslant 0$.

An uneasy feeling should prevail at this point because the arbitrary addition of terms to equations is not orthodox mathematics. Further treatment of the

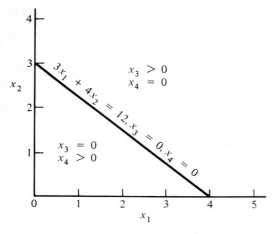

Figure 12-6 Slack and artificial variables.

artificial variable is necessary, and this treatment will be explained as part of Example 12-2.

## 12-16 SIMPLEX ALGORITHM APPLIED TO MINIMIZATION WITH GREATER-THAN CONSTRAINTS

**EXAMPLE 12-2** Determine the minimum value of $y$ and the magnitudes of $x_1$ and $x_2$ at this minimum, where

$$y = 6x_1 + 3x_2$$

subject to the constraints

$$5x_1 + x_2 \geqslant 10$$
$$9x_1 + 13x_2 \geqslant 74$$
$$x_1 + 3x_2 \geqslant 9$$

SOLUTION  Because this problem involves only two physical variables, the constraints and lines of constant $y$ can be graphed, as in Fig. 12-7.

For the solution by linear programming, first write the constraint inequalities as equations by introducing the slack variables $x_3$, $x_4$, and $x_5$.

$$5x_1 + x_2 - x_3 = 10$$
$$9x_1 + 13x_2 - x_4 = 74$$
$$x_1 + 3x_2 - x_5 = 9$$

Next the artificial variables, $x_6$, $x_7$, and $x_8$, are inserted in each equation

$$5x_1 + x_2 - x_3 + x_6 = 10 \qquad (12\text{-}23)$$

$$9x_1 + 13x_2 \quad - x_4 \qquad + x_7 \quad = 74 \qquad (12\text{-}24)$$

$$x_1 + 3x_2 \qquad - x_5 \qquad + x_8 = 9 \qquad (12\text{-}25)$$

Now the remaining conditions surrounding the artificial variables are specified in that the objective function is also revised as follows:

$$y = 6x_1 + 3x_2 + Px_6 + Px_7 + Px_8 \qquad (12\text{-}26)$$

The coefficient $P$ never assumes a numerical value but is only considered to be extremely large. The existence of the products of $P$ and the artificial variables in Eq. (12-26) is so penalizing to the minimization attempt that no satisfactory minimization will occur until the values of the artificial variables have been driven to zero.

The linear-programming process can now start at the origin in Fig. 12-7 with all the slack variables equal to zero but all the artificial variables having positive values. The simplex algorithm applies, but one further operation is performed on the objective function before writing the first tableau. From Eqs. (12-23) to (12-25)

$$x_6 = 10 - 5x_1 - x_2 + x_3$$
$$x_7 = 74 - 9x_1 - 13x_2 \quad + x_4$$
$$x_8 = 9 - x_1 - 3x_2 \qquad + x_5$$

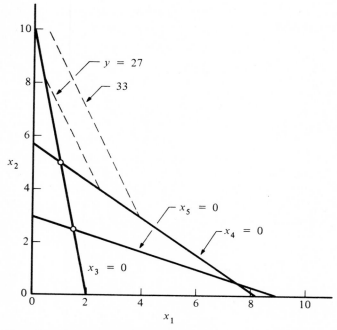

Figure 12-7 Minimization in Example 12-2.

**Table 12-8 Tableau 1 of Example 12-2**

| | $x_1 = 0$ | $x_2 = 0$ | $x_3 = 0$ | $x_4 = 0$ | $x_5 = 0$ | $x_6$ | $x_7$ | $x_8$ | |
|---|---|---|---|---|---|---|---|---|---|
| 10 | 5 | 1 | -1 | | | 1 | | | 10 |
| 74/13 | 9 | 13 | | -1 | | | 1 | | 74 |
| ⟹ 3 | 1 | 3 | | | -1 | | | 1 | 9 |
| | $15P - 6$ | $17P - 3$ | $-P$ | $-P$ | $-P$ | | | | $93P$ |

Substituting these values of the artificial variables into Eq. (12-26) and grouping gives

$$y = (6 - 15P)x_1 + (3 - 17P)x_2 + Px_3 + Px_4 + Px_5 + 93P$$

The first tableau can now be constructed. In the minimization operation the variable with the *largest positive* difference coefficient is chosen to be programmed. In tableau 1 (Table 12-8) this choice will be $x_2$ because $P$ appears with the largest coefficient and the $P$ values dominate over any purely numerical terms.

The next question is: What is the limit to which $x_2$ can be increased? Just as in the maximization process, the variable $x_2$ is increased to the most limiting constraint, which in this case is the third one. The procedure in transforming to tableau 2 (Table 12-9) is the standard simplex algorithm. The difference coefficient with the largest positive value is $(28P - 15)/3$, so $x_1$ is programmed next until limited by the point where $x_6$ starts to go negative in the first constraint. In tableau 3 (Table 12-10) $x_5$ has the largest positive difference coefficient, so $x_5$ is increased until it reaches the limiting value of 7, beyond which $x_7$ would become negative.

**Table 12-9 Tableau 2 of Example 12-2**

| | $x_1 = 0$ | $x_2$ | $x_3 = 0$ | $x_4 = 0$ | $x_5 = 0$ | $x_6$ | $x_7$ | $x_8 = 0$ | |
|---|---|---|---|---|---|---|---|---|---|
| ⟹ 3/2 | 14/3 | 0 | -1 | 0 | 1/3 | 1 | 0 | -1/3 | 7 |
| 105/14 | 14/3 | 0 | 0 | -1 | 13/3 | 0 | 1 | -13/3 | 35 |
| 9 | 1/3 | 1 | 0 | 0 | -1/3 | 0 | 0 | 1/3 | 3 |
| | $\dfrac{28P - 15}{3}$ | 0 | $-P$ | $-P$ | $\dfrac{14P - 3}{3}$ | 0 | 0 | $\dfrac{-17P + 3}{3}$ | $42P + 9$ |

**Table 12-10  Tableau 3 of Example 12-2**

|  | $x_1$ | $x_2$ | $x_3 = 0$ | $x_4 = 0$ | $x_5 = 0$ | $x_6 = 0$ | $x_7$ | $x_8 = 0$ |  |
|---|---|---|---|---|---|---|---|---|---|
| 21 | 1 | 0 | $-3/14$ | 0 | $1/14$ | $3/14$ | 0 | $-1/14$ | $3/2$ |
| $\Longrightarrow 7$ | 0 | 0 | 1 | $-1$ | 4 | $-1$ | 1 | $-4$ | 28 |
| $-7$ | 0 | 1 | $1/14$ | 0 | $-5/14$ | $-1/14$ | 0 | $5/14$ | $5/2$ |
|  | 0 | 0 | $\dfrac{14P - 15}{14}$ | $-P$ | $\dfrac{56P - 9}{14}$ | $\dfrac{-28P + 15}{14}$ | 0 | $\dfrac{-70P + 9}{14}$ | $\dfrac{56P + 33}{2}$ |

The arrow ⇓ points to the $x_4 = 0$ column.

**Table 12-11  Tableau 4 of Example 12-2**

| $x_1$ | $x_2$ | $x_3 = 0$ | $x_4 = 0$ | $x_5$ | $x_6 = 0$ | $x_7 = 0$ | $x_8 = 0$ |  |
|---|---|---|---|---|---|---|---|---|
| 1 | 0 | $-13/56$ | $1/56$ | 0 | $13/56$ | $-1/56$ | 0 | 1 |
| 0 | 0 | $1/4$ | $-1/4$ | 1 | $-1/4$ | $1/4$ | $-1$ | 7 |
| 0 | 1 | $9/56$ | $-5/56$ | 0 | $-9/56$ | $5/56$ | 0 | 5 |
| 0 | 0 | $-51/56$ | $-9/56$ | 0 | $-P + 51/56$ | $-P + 9/51$ | $-P$ | 21 |

In tableau 4 (Table 12-11) all difference coefficients are negative, so no further reduction in the objective function is possible. The solution is

$$x_1^* = 1 \quad \text{and} \quad x_2^* = 5$$

at which point $y^* = 21$.

## 12-17  REVIEW OF MINIMIZATION CALCULATION

Now that the minimization process in Example 12-2 has been completed, a re-examination of the successive tableaux in the problem will present a more complete picture of the operation. The introduction of the artificial variables in the objective function and the constraint equations permits a temporary violation of the constraints but only at the expense of an enormously large value of the objective function. The solution will certainly not be satisfactory until all the $P$ terms are removed from the expression for the objective function.

In contrast to the maximization problem, e.g., Example 12-1, where the constraints were like solid walls that could not be surmounted, in the minimization problem the constraints are like stiff rubber bands that can be violated temporarily, but with a severe penalty in the magnitude of the objective function.

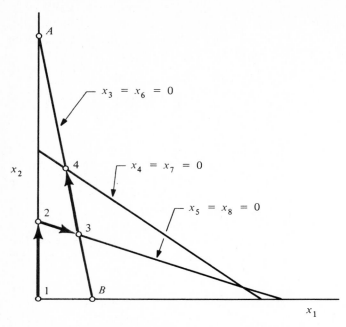

**Figure 12-8** Points represented by successive tableaux in Example 12-2.

Looking first at tableau 1, we see that the position represented in Fig. 12-8 is the origin, because $x_1 = x_2 = 0$. The slack variables $x_3$, $x_4$, and $x_5$ are also zero, but the artificial variables have nonzero values: $x_6 = 10$, $x_7 = 74$, and $x_8 = 9$. The value of the objective function is $93P$, which is prohibitively large.

In tableau 2 the nonzero values of variables are $x_2 = 3$, $x_6 = 7$, and $x_7 = 35$. In moving from point 1 to point 2 in Fig. 12-8, the magnitude of the objective function drops from $93P$ to $42P + 9$. In other words, the pure numerical value increases by 9, but the staggering $P$ term decreases from $93P$ to $42P$.

The next shift is to point 3, represented by tableau 3, where the nonzero values are $x_1 = \frac{3}{2}$, $x_2 = \frac{5}{2}$, and $x_7 = 28$. The magnitude of the objective function is $(56P + 33)/2$, showing a continued increase of the numerical portion and a decrease of the $P$ coefficient. At this point $x_3$ and $x_6$ are zero, as are $x_5$ and $x_8$. Point 3 requires a positive value of the $-x_4 + x_7$ combination which is achieved by a zero value of $x_4$ and a positive value of $x_7$.

The final move is from point 3 to point 4, where the nonzero values are $x_1 = 1$, $x_2 = 5$, and the slack variable $x_5 = 7$. All the artificial variables are now zero, so the objective function in tableau 4 is free of $P$ terms and has the value of 21. Since there are no remaining positive difference coefficients, the minimum value has been reached.

The minimization process thus consisted of starting at the origin, which was made permissible but highly uncomfortable by the introduction of artificial variables, and then moving through the thicket of constraint lines until finally the light of day was seen at point 4. In this problem, the first feasible point encountered was also the minimum, but it is possible for the process to move around several corners after it breaks into the feasible region.

## 12-18 APPEARANCE OF NEGATIVE NUMBER FOR MOST RESTRICTIVE CONSTRAINT

One further comment should be made regarding tableau 3, where, following the decision to program $x_5$, the limitation for increasing $x_5$ set by the first equation was 21, by the second equation 7, and by the third equation $-7$. Strict consistency would have required us to choose the third as the controlling equation. Geometrically, in Fig. 12-8, the test to find the controlling equation consists of moving along the $-x_3 + x_6 = 0$ line. Beyond point $A$, where $x_5 = 21$, the value of $x_1$ would go negative. Beyond 4, where $x_5 = 7$, $x_7$ would go negative. The limitation of $x_5 = -7$ is represented by point $B$, which would result in a step backward, so the limitations with negative values are ignored.

## 12-19 EQUALITIES AND SENSES OF INEQUALITIES

The two classes of linear-programming problems considered so far are (1) maximization with less-than constraints and (2) minimization with greater-than constraints. Physical problems sometimes require mixtures of senses of the constraints, i.e., some less-than and the remainder greater-than constraints. One or more equality constraints may also appear in combination with the inequalities. The next several sections explore these combinations and certain other situations introduced by such physical requirements as mass balances.

## 12-20 MIXTURES OF EQUALITIES AND INEQUALITIES AND OF SENSES OF INEQUALITIES

A few general guidelines are applicable to setting up problems where the constraints are mixed. These guidelines control the introduction of slack and artificial variables. In the two classes of problems treated so far, where the constraints were all inequalities with the same sense, the insertion of slack and artificial variables accomplished two functions: (1) converting all inequalities into equalities and (2) permitting programming to begin at the origin of the physical variables. The same objective prevails if the constraints are mixed. One additional variation is that when an artificial variable is introduced, it must appear in the objective function as $+Px_i$ if the objective function is being minimized and $-Px_i$ if the function is being maximized.

**Example 12-3** Find the maximum of $y$, where

$$y = 3x_1 + 2x_2 + 4x_3$$

subject to

$$3x_1 + 4x_2 + 5x_3 \leqslant 40 \qquad (12\text{-}27)$$

$$x_1 + x_2 + x_3 = 9 \qquad (12\text{-}28)$$

$$7x_1 + 4x_2 + 4x_3 \geqslant 42 \qquad (12\text{-}29)$$

SOLUTION Equation (12-27) requires only the insertion of a positive slack variable $x_4$ in order to convert the inequality into an equality and also permit $x_1 = x_2 = x_3 = 0$ with $x_4 \geqslant 0$. Equation (12-27) thus becomes

$$3x_1 + 4x_2 + 5x_3 + x_4 = 40$$

Equation (12-28) is already an equality, but the equation will not permit programming to start at the origin. An artificial variable $x_5$ will be introduced into Eq. (12-28) which retains the equality and also permits $x_1 = x_2 = x_3 = 0$,

$$x_1 + x_2 + x_3 + x_5 = 9 \qquad (12\text{-}30)$$

The objective function must also be revised so that $x_5$ will ultimately be driven to zero to satisfy Eq. (12-28),

$$y = 3x_1 + 2x_2 + 4x_3 - Px_5 \qquad (12\text{-}31)$$

The penalty term in the objective function is assigned a negative value because the goal is to maximize the function and the maximum should not be reached until $x_5$ has shrunk to zero.

The final constraint, Eq. (12-29), requires both a slack and an artificial variable. The slack variable $-x_6$ converts the inequality into an equality. The artificial variable $x_7$ allows programming to start at the origin. Equation (12-29) then becomes

$$7x_1 + 4x_2 + 4x_3 - x_6 + x_7 = 42 \qquad (12\text{-}32)$$

Introducing the penalty term associated with $x_7$ into the objective function revises Eq. (12-31) further

$$y = 3x_1 + 2x_2 + 4x_3 - Px_5 - Px_7 \qquad (12\text{-}33)$$

As the final step before writing the first tableau, substitute $x_5$ from Eq. (12-30) and $x_7$ from Eq. (12-32) into the objective function, Eq. (12-33), to get

$$y = -51P + (8P + 3)x_1 + (5P + 2)x_2 + (5P + 4)x_3 - Px_6$$

### Table 12-12  Tableau 1 of Example 12-3

|        | $x_1 = 0$ | $x_2 = 0$ | $x_3 = 0$ | $x_4$ | $x_5$ | $x_6 = 0$ | $x_7$ |        |
|--------|-----------|-----------|-----------|-------|-------|-----------|-------|--------|
| 40/3   | 3         | 4         | 5         | 1     |       |           |       | 40     |
| 9      | 1         | 1         | 1         |       | 1     |           |       | 9      |
| $\Longrightarrow$ 6 | 7 | 4 | 4 |       |       | $-1$      | 1     | 42     |
|        | $-8P - 3$ | $-5P - 2$ | $-5P - 4$ |       |       | $P$       |       | $-51P$ |

The first tableau is shown in Table 12-12. Since programming begins at the origin, $x_1$, $x_2$, and $x_3$ are assigned the value of zero. The slack variable $x_6$ is also zero because the nonzero value of $x_7$ permits $x_1$, $x_2$, and $x_3$ in Eq. (12-32) to go to zero.

Since the goal is the maximum, the variable with the largest negative difference coefficient, which is $x_1$, will be programmed first. The succession of tableaux is shown in Tables 12-13 to 12-16.

### Table 12-13  Tableau 2 of Example 12-3

|        | $x_1$ | $x_2 = 0$ | $x_3 = 0$ | $x_4$ | $x_5$ | $x_6 = 0$ | $x_7 = 0$ |         |
|--------|-------|-----------|-----------|-------|-------|-----------|-----------|---------|
| $\Longrightarrow$ 154/23 | 0 | 16/7 | 23/7 | 1 | 0 | 3/7 | $-3/7$ | 22 |
| 7      | 0     | 3/7       | 3/7       | 0     | 1     | 1/7       | $-1/7$    | 3       |
| 42/4   | 1     | 4/7       | 4/7       | 0     | 0     | $-1/7$    | 1/7       | 6       |
|        | 0     | $\dfrac{-3P - 2}{7}$ | $\dfrac{-3P - 16}{7}$ | 0 | 0 | $\dfrac{-P - 3}{7}$ | $\dfrac{8P + 3}{7}$ | $-3P + 18$ |

### Table 12-14  Tableau 3 of Example 12-3

|        | $x_1$ | $x_2 = 0$ | $x_3$ | $x_4 = 0$ | $x_5$ | $x_6 = 0$ | $x_7 = 0$ |          |
|--------|-------|-----------|-------|-----------|-------|-----------|-----------|----------|
| 154/16 | 0     | 16/23     | 1     | 7/23      | 0     | 3/23      | $-3/23$   | 154/23   |
| $\Longrightarrow$ 1 | 0 | 3/23 | 0 | $-3/23$ | 1 | 2/23 | $-2/23$ | 3/23 |
| 25/2   | 1     | 4/23      | 0     | $-4/23$   | 0     | $-5/23$   | 5/23      | 50/23    |
|        | 0     | $\dfrac{-3P + 30}{23}$ | 0 | $\dfrac{3P + 16}{23}$ | 0 | $\dfrac{-2P - 3}{23}$ | $\dfrac{25P + 3}{23}$ | $\dfrac{-3P + 766}{23}$ |

**Table 12-15  Tableau 4 of Example 12-3**

|  | $x_1$ | $x_2$ | $x_3$ | $x_4 = 0$ | $x_5 = 0$ | $x_6 = 0$ | $x_7 = 0$ |  |
|---|---|---|---|---|---|---|---|---|
| $-18$ | 0 | 0 | 1 | 1 | $-16/3$ | $-1/3$ | $1/3$ | 6 |
| $\Rightarrow 3/2$ | 0 | 1 | 0 | $-1$ | $23/3$ | $2/3$ | $-2/3$ | 1 |
| $-6$ | 1 | 0 | 0 | 0 | $-4/3$ | $-1/3$ | $1/3$ | 2 |
|  | 0 | 0 | 0 | 2 | $\dfrac{3P - 30}{3}$ | $-1$ | $P + 1$ | 32 |

**Table 12-16  Tableau 5 of Example 12-3**

| $x_1$ | $x_2 = 0$ | $x_3$ | $x_4 = 0$ | $x_5 = 0$ | $x_6$ | $x_7 = 0$ |  |
|---|---|---|---|---|---|---|---|
| 0 | $1/2$ | 1 | $1/2$ | $-3/2$ | 0 | 0 | $13/2$ |
| 0 | $3/2$ | 0 | $-3/2$ | $23/2$ | 1 | $-1$ | $3/2$ |
| 1 | $1/2$ | 0 | $-1/2$ | $5/2$ | 0 | 0 | $5/2$ |
| 0 | $3/2$ | 0 | $1/2$ | $\dfrac{2P + 3}{2}$ | 0 | $P$ | 33.5 |

The $P$ terms have disappeared from the difference coefficients associated with the physical variables $x_1$, $x_2$, and $x_3$, and so the position indicated by tableau 4 is now in the feasible region. The optimal solution, then, is $x_1^* = 2.5$, $x_2^* = 0$, $x_3^* = 6.5$, and $y^* = 33.5$.

## 12-21  MATERIAL BALANCES AS CONSTRAINTS

In the application of optimization methods, including linear programming, to thermal systems, material balances often impose constraints. The input-output material balance often results in a unique situation during the simplex procedure that should be recognized in order to provide the interpretation that allows the procedure to continue. The situation will be illustrated by an example.

**Example 12-4**  In the processing plant shown in Fig. 12-9 the operation is essentially one of concentrating material A. The concentrator receives a raw material consisting of 40 percent A by mass and can supply two products of 60 and 80 percent A, respectively. The flow rate of the raw material is des-

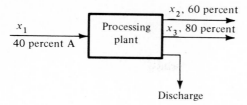

**Figure 12-9** Concentrator in Example 12-4.

ignated $x_1$ metric tons per day and the 60 percent and 80 percent products are designated $x_2$ and $x_3$, respectively. The prices are:

| Amount | $x_1$ | $x_2$ | $x_3$ |
|---|---|---|---|
| Price per metric ton | $40 | $80 | $120 |

The capacity of the loading facility imposes the constraint

$$2x_2 + 3x_3 \leqslant 60$$

Determine the combination of raw material and products that results in maximum profit for the plant.

SOLUTION The objective function is the difference between the income of the products and the cost of the raw materials

$$y = 80x_2 + 120x_3 - 40x_1$$

In addition to the constraint of the loading facility, the mass balance of material A imposes another constraint that can be expressed by either Eq. (12-34) or (12-35)

$$0.6x_2 + 0.8x_3 = 0.4x_1 \tag{12-34}$$

$$0.6x_2 + 0.8x_3 \leqslant 0.4x_1 \tag{12-35}$$

Equation (12-34) is a strict equality for material, while Eq. (12-35) allows for the possibility of dumping some of material A during the processing. In some rare instances the most economical solution can be achieved by dumping some material of value, so in the solution presented below the possibility of dumping will be used and Eq. (12-35) will be chosen. The mathematical statement of the problem becomes:

Maximize $$y = 80x_2 + 120x_3 - 40x_1$$

subject to $$2x_2 + 3x_3 \leqslant 60$$

$$-0.4x_1 + 0.6x_2 + 0.8x_3 \leqslant 0$$

The first tableau is shown in Table 12-17. The variable to be programmed

**Table 12-17  Tableau 1 of Example 12-4**

| | $x_1 = 0$ | $x_2 = 0$ | $x_3 = 0$ | $x_4$ | $x_5$ | |
|---|---|---|---|---|---|---|
| 20 | 0 | 2 | 3 | 1 | | 60 |
| $\Longrightarrow 0/0.8 = 0$ | −0.4 | 0.6 | 0.8 | | 1 | 0 |
| | 40 | −80 | −120 | | | 0 |

**Table 12-18  Tableau 2 of Example 12-4**

| | $x_1 = 0$ | $x_2 = 0$ | $x_3$ | $x_4$ | $x_5 = 0$ | |
|---|---|---|---|---|---|---|
| $\Longrightarrow 40$ | 1.5 | −0.25 | 0 | 1 | −3.75 | 60 |
| $0/-0.5 = -0$ | −0.5 | 0.75 | 1 | 0 | 1.25 | 0 |
| | −20 | 10 | 0 | 0 | 150 | 0 |

first is $x_3$ for which the most restrictive constraint is the second one. Using the simplex algorithm gives tableau 2 (Table 12-18).

The difference coefficient of $x_1$ in tableau 2 has the largest negative value, so $x_1$ is programmed next. The straightforward procedure would indicate that the second constraint controls, because $0 < 40$. The key feature in solving this class of problems, however, is the interpretation of the indicators of control. We shall interpret the zero as a "negative zero" and, as is the practice, ignore any constraint associated with a negative number in the left column. The first constraint is used as the controlling one and results in tableau 3 (Table 12-19). No negative difference coefficients remain, so the maximum has been reached: $x_1^* = 40$, $x_3^* = 20$, and $y^* = 800$.

**Table 12-19  Tableau 3 of Example 12-4**

| $x_1$ | $x_2 = 0$ | $x_3$ | $x_4 = 0$ | $x_5 = 0$ | |
|---|---|---|---|---|---|
| 1 | −0.1667 | 0 | 0.6667 | −2.5 | 40 |
| 0 | 0.833 | 1 | 0.333 | 0 | 20 |
| 0 | 13.33 | 0 | 13.33 | 100 | 800 |

**Table 12-20  Incorrect Tableau 3**

| $x_1$ | $x_2 = 0$ | $x_3 = 0$ | $x_4$ | $x_5 = 0$ | |
|---|---|---|---|---|---|
| 0 | 2 | 3 | 1 | 0 | 60 |
| 1 | −1.5 | −2 | 0 | −2.5 | 0 |
| 0 | −10 | −40 | 0 | 100 | 0 |

Let us now return to tableau 2 and explore what would have occurred if the second equation had been chosen as the controlling one. That tableau would have been as shown in Table 12-20. The values indicated by the tableau in Table 12-20 are $x_1 = x_2 = x_3 = x_5 = 0$, and $x_4 = 60$. These are precisely the starting conditions of tableau 1, so the process is in an infinite loop. The interpretation of the negative zero breaks open the loop.

The physical explanation of what occurs between tableaux 1 and 2 is as follows. Most profit can be made by selling $x_3$, so this variable is understandably the one to be increased first. But tableau 2 shows that $x_3$ is still zero, and the reason is that the second constraint, the material balance, will not allow $x_3$ to increase until some raw material $x_1$ has also been brought in. By the shift of the slack variable $x_5$ to zero in the column heading, the next programming operation will assure that $x_1$ is increased in proper proportion to the product $x_3$.

## 12-22  FURTHER TOPICS IN LINEAR PROGRAMMING

This chapter is an introduction to linear programming but explains enough about the procedures to solve some practical optimizations of thermal systems. All the examples and the problems that follow can be solved by hand calculations. The engineer first encountering linear programming probably should do some hand calculations to obtain insight into the methods. Thereafter one of the many canned programs available in most computer centers should be used to avoid the tedium of the calculations.

A valuable contribution that linear programming can make is in *sensitivity analysis*, where following the solution of the optimization an individual coefficient in the constraint equations or in the objective function is analyzed to determine what influence a slight change in it will have on the optimal value. An application of sensitivity analysis might be to determine what the influence on the peak operating effectiveness of a processing plant might be if one certain heat exchanger could be enlarged.

This chapter concentrated on techniques, but as with the other optimization methods presented in this book, once some familiarity with the technique has been achieved, the challenge is (1) to recognize that a possibility for optimization exists and (2) to set up the problem in a mathematical form permitting linear programming to be applied.

## PROBLEMS

12-1 A farmer wishes to choose the proportions of corn and soybeans to plant to achieve maximum net return. He has 240 hectares available for planting although he may elect to operate less than 240 hectares. The capital and labor requirements and the net return on each of the two crops are

| Crop | Capital cost per hectare | Labor, worker-hours per hectare | Net return per hectare |
|------|--------------------------|---------------------------------|------------------------|
| Corn | $120 | 5 | $50 |
| Soybeans | 40 | 10 | 75 |

The farmer has a maximum of 2000 worker-hours of labor available and maximum capital of $19,200.

(a) Set up the objective function and constraints.

(b) Use the simplex algorithm of linear programming to determine the optimal number of hectares of corn and soybeans and the maximum net return.

Ans.: Maximum net return = $16,000.

12-2 During a summer session a student enrolls in two courses, psychology for four hours and engineering for three hours. The student wants to use the available time to amass the largest number of grade points (numerical grade multiplied by the number of hours). Students who have taken these courses previously indicate that probable grades are functions of the time spent:

$$G_e = \frac{x_1}{5} \quad \text{and} \quad G_p = \frac{x_2}{8}$$

where $G$ = numerical grade (4.0 = A, 3.0 = B, etc.)

$x_1$ = number of hours per week spent outside of class studying engineering

$x_2$ = number of hours per week spent outside of class studying psychology

The total number of hours available for outside study per week cannot exceed 39. Furthermore, most engineering students can tolerate the two courses in a combination such that

$$x_1 + 0.5x_2 \leqslant 27$$

Also it is fruitless to spend more time studying beyond that necessary to earn an A.

Use the simplex algorithm to determine how to distribute study time, subject to the constraints, in order to acquire the maximum number of grade points.

Ans.: B in both courses.

12-3 A food-processing firm is planning construction of a plant that could manufacture a combination of three frozen food products, pot pies, TV dinners, and pizzas. The investment costs consist of plant costs and machinery costs, and the credit rating of the firm will permit loans of $4.8 million or less for building construction and $1.6 million or less for machinery. The building and machinery costs for each of the proposed products are

| Food product | Building cost | Machinery cost |
|---|---|---|
| Pot pies | $600,000$x_1$ | $100,000$x_1$ |
| TV dinners | 500,000$x_2$ | 400,000$x_2$ |
| Pizzas | 400,000$x_3$ | 200,000$x_3$ |

where $x_1$, $x_2$, and $x_3$ represent the hourly production rate of pot pies, TV dinners, and pizzas, respectively, in thousands of units.

The hourly profit from the manufacture of each of the products in dollars is $25x_1$, $30x_2$, and $40x_3$, respectively.

The values of $x_1$, $x_2$, and $x_3$ are to be determined such that the profit is maximum within the prevailing constraints.

(*a*) Set up the objective function and constraints.

(*b*) Use the simplex algorithm of linear programming to determine optimal values of $x_1, x_2, x_3$, and the hourly profit.

**Ans.:** Optimum profit = $340 per hour.

**12-4** In the manufacture of cement the basic operations are to grind limestone, mix with clay or shale, and then heat the mixture in a rotary kiln, as shown in Fig. 12-10. A certain cement plant can produce three ASTM types of cement:

*Type I.*   Standard portland
*Type II.*  Sulfate- and alkaline-resistant
*Type III.* High early strength

The profit of each type and the capabilities of the grinder and kiln in processing these cements are shown in the table.

| Cement type | Profit, per megagram | Grinder capacity, Gg/day | Kiln capacity Gg/day |
|---|---|---|---|
| I | $ 6 | Coarse 10 | 8 |
| II | 10 | Fine   5 | 4.8 |
| III | 9 | Fine   5 | 6 |

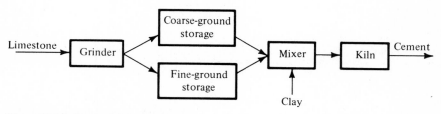

**Figure 12-10** Cement plant in Prob. 12-4.

The grinder capacity shown in the table of 10 Gg/day for type I means, for example, that the grinder could grind the limestone for 10 Gg of type I if it operated all day solely on limestone for type I. The grinder and kiln operate 24 h/day and can switch from one cement type to another instantaneously. The limestone storage space and mixer capacity are more than adequate for any rates that the grinder and kiln will permit.

Use the simplex algorithm of linear programming to determine what daily production of the various types of cement will result in maximum profit.

**Ans.:** Maximum daily profit = $51,000.

**12-5** Three materials $A$, $B$, and $C$ of varying thicknesses are available for combining into a building wall, as shown in Fig. 12-11. The characteristics and costs of the materials are

| Material | Thermal resistance, units per centimeter thickness | Load-bearing capacity, units/cm | Cost per centimeter |
|---|---|---|---|
| $A$ | 30 | 7 | $8 |
| $B$ | 20 | 2 | 4 |
| $C$ | 10 | 6 | 3 |

The total thermal resistance of the wall must be 120 or greater, and the total load-bearing capacity must be 42 or greater. The minimum-cost wall is sought.

(a) Set up the objective function and constraints.

(b) Use the simplex algorithm of linear programming to determine the optimal thicknesses of each material.

**Ans.:** $30 minimum cost.

**12-6** The optimization of the combined gas- and steam-turbine plant in Prob. 7-4 resulted in a linear objective function and three linear constraints. Use the simplex algorithm to determine the optimum value of $q_1$ and $q_2$. To simplify mathematical manipulation, use the following equations instead of those in Prob. 7-4:

Objective function: $$q = q_1 + q_2$$

subject to $$q_1 + 1.2q_2 \geqslant 28 \text{ MW}$$

$$q_1 + 0.4q_2 \geqslant 19 \text{ MW}$$

$$q_1 + 1.7q_2 \geqslant 32 \text{ MW}$$

**Ans.:** Optimal $q$ = 25.75 MW.

**12-7** The furnace serving a certain steam-generating plant is capable of burning coal, oil, and gas simultaneously. The heat-release rate of the furnace must be 2400 kW, which with the 75 percent combustion efficiency of this furnace requires a combined thermal input rate in the fuel of 3200 kW. Ordinances in certain cities impose a limit on the average sulfur con-

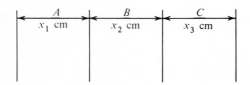

**Figure 12-11** Composite wall in Prob. 12-5.

tent of the fuel mixture, and in the city where this plant is located the limit is 2 percent or less. The sulfur contents, costs, and heating values of the fuels are shown in the table.

| Fuel | Sulfur content, % | Cost per megagram | Heating value, kJ/kg |
|------|------|------|------|
| Coal | 3.0 | $24 | 35,000 |
| Oil | 0.4 | 36 | 42,000 |
| Gas | 0.2 | 42 | 55,000 |

Using the simplex algorithm of linear programming, determine the combination of fuel rates that results in minimum costs and yet meets all constraints.

**Ans.:** Minimum cost = $0.00231 per second.

**12-8** A manufacturer of cattle food mixes a combination of wheat and soybeans to form a product which has minimum requirements of 24 percent protein and 1.2 percent minerals by mass. The composition and prices of the wheat and soybeans are given in the table.

| Constituent | Percent composition by mass | | Price per 100 kg |
|------|------|------|------|
| | Protein | Minerals | |
| Wheat | 16 | 1.5 | $10 |
| Soybeans | 48 | 1.0 | 20 |

Use the simplex algorithm of linear programming to determine the mass of each of the constituents in 100 kg of product such that the cost of the raw materials is minimum and the nutritional requirements are met.

**Ans.:** 75 kg of wheat and 25 kg of soybeans.

**12-9** A wax concentrating plant (Fig. 12-12) receives feedstock with a low concentration of wax and refines it into a product with a high concentration of wax. The selling prices of the products are $x_1$, $8 per megagram and $x_2$, $6 per megagram. The raw material costs are $x_3$, $1.5 per megagram and $x_4$, $3 per megagram. The plant operates under the following constraints:

1. No more wax leaves the plant than enters.
2. The receiving facilities of the plant are limited to a total of 1600 Mg/h.
3. The packaging facilities can accommodate a maximum of 1200 Mg/h of $x_2$ or 1000 Mg/h of $x_1$ and can switch from one to the other with no loss of time.

If the operating cost of the plant is constant, use the simplex algorithm of linear pro-

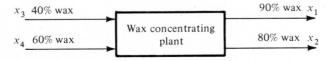

**Figure 12-12** Wax concentration plant in Prob. 12-9.

gramming to determine the purchase and production plan that results in the maximum profit.

**Ans.:** Profit = $3650 per hour.

**12-10** A dairy operating on the flow diagram shown in Fig. 12-13 can buy raw milk from either or both of two sources and can produce skim milk, homogenized milk, and half-and-half cream. The costs and butterfat contents of the sources and products are

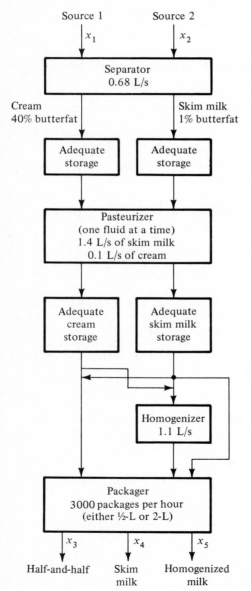

**Figure 12-13** Dairy in Prob. 12-10.

| Item | Designation, L/day | Sale or purchase cost | Butterfat content, vol % |
|---|---|---|---|
| Source 1 | $x_1$ | $0.23 per liter | 4.0 |
| Source 2 | $x_2$ | 0.24 per liter | 4.5 |
| Half-and-half | $x_3$ | 0.48 per half-liter | $\geqslant 10.0$ |
| Skim milk | $x_4$ | 0.60 per two-liter | $\geqslant 1.0$ |
| Homogenized milk | $x_5$ | 0.68 per two-liter | $\geqslant 3.0$ |

The daily quantities of sources and products are to be determined so that the plant operates with maximum profit. All units in the dairy can operate for a maximum of 8 h/day.

(a) Write the equation for the profit and the constraint equations in terms of the $x$ variables.

(b) Use linear programming to solve for the plan that results in maximum profit.

**Ans.:** (a)

$$y = -0.23x_1 - 0.24x_2 + 0.96x_3 + 0.30x_4 + 0.34x_5$$

subject to

Separator:
$$x_1 + x_2 \leqslant 19{,}584$$

Pasteurizer:
$$x_1 + 1.08x_2 \leqslant 20{,}150$$

Homogenizer:
$$x_5 \leqslant 31{,}680$$

Packager:
$$4x_3 + x_4 + x_5 \leqslant 48{,}000$$

Butterfat:
$$-4x_1 - 4.5x_2 + 10x_3 + x_4 + 3x_5 \leqslant 0$$

Total mass:
$$-x_1 - x_2 + x_3 + x_4 + x_5 \leqslant 0$$

(b) Maximum profit = $5908 per day.

**12-11** A chemical plant whose flow diagram is shown in Fig. 12-14 manufactures ammonia, hydrochloric acid, urea, ammonium carbonate, and ammonium chloride from carbon dioxide, nitrogen, hydrogen, and chlorine. The $x$ values in Fig. 12-14 indicate flow rates in moles per second.

The costs of the feed stocks are $c_1$, $c_2$, $c_3$, and $c_4$ dollars per mole, and the values of the products are $p_5$, $p_6$, $p_7$, and $p_8$ dollars per mole, where the subscript corresponds to that of the $x$ value. In reactor 3 the ratios of molal flow rates are $m = 3x_7$ and $x_1 = 2x_7$, and in the other reactors straightforward material balances apply. The capacity of reactor 1 is equal to or less than 2 mol/s of $NH_3$, and the capacity of reactor 2 is equal to or less than 1.5 mol/s.

(a) Develop the expression for the profit.

(b) Write the constraint equations for this plant.

**Ans.:** (b)

$$x_1 - 2x_7 = 0$$

$$2x_2 - x_5 - 3x_7 - x_8 = 0$$

$$2x_3 - 3x_5 - x_6 - 9x_7 - 4x_8 = 0$$

$$2x_4 - x_6 - x_8 = 0$$

$$x_6 + x_8 \leqslant 1.5$$

$$x_5 + 3x_7 + x_8 \leqslant 2.0$$

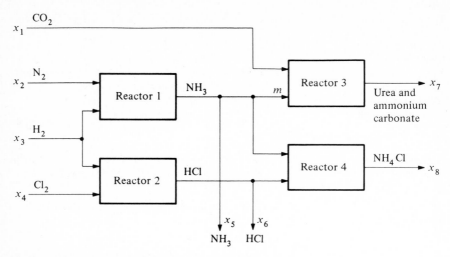

Figure 12-14 Flow diagram of chemical plant in Prob. 12-11.

12-12 When large fabric filter installations, called *baghouse filters*, filter high-temperature gases, e.g., from a smelter, the temperature of the gases must be reduced to 265°C or less, even when using glass-fiber filters. Three methods[1] of reducing the temperature may be used singly or in combination, as shown in Fig. 12-15: (1) reject heat to ambient air through the use of a heat exchanger, (2) dilute the hot gas with ambient air that is at a temperature of 25°C, and (3) inject water for evaporative cooling. If either dilution air or water injection is used, the baghouse must be enlarged to accommodate the additional mass flow. Designate

$$x_1 = \text{area of heat exchanger, m}^2$$

$$x_2 = \text{mass flow rate of dilution air, kg/s}$$

$$x_3 = \text{mass flow rate of spray water, kg/s}$$

*Data*

Cost of baghouse is $2000 for each kilogram per second of capacity.
The heat-transfer surface costs $15 per square meter.

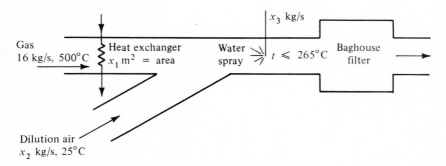

Figure 12-15 Cooling a gas before it enters a baghouse filter.

There is no cost for the dilution air and spray water other than that due to the enlargement of the baghouse.

The entering flow rate of gas = 16 kg/s at 500°C.

Each square meter of heat exchanger reduces the temperature of gas by 2°C.

The evaporation of spray water cools the gas-air mixture equivalent to a sensible-heat removal from the gas entering the water spray of 2000 kJ/kg of spray water.

The specific heat of the gas, dilution air, and gas-air mixture is 1.0 kJ/(kg · K).

To avoid corrosion, the mass rate of flow of spray water must be 5 percent or less of the combined flow of gas and dilution air.

(a) Set up the mathematical statement of the optimization problem in terms of $x_1$ to $x_3$ to minimize the cost, subject to the appropriate constraints.

(b) Develop the first tableau of the simplex algorithm of the linear-programming solution of this optimization and indicate which variable to program first.

**Ans.:** (a) $y = 32,000 + 15x_1 + 2000x_2 + 2000x_3$

subject to
$$2x_1 + 15x_2 + 125x_3 \geqslant 235$$
$$-x_2 + 20x_3 \leqslant 16$$

**12-13** Some petrochemical plants take a large flow rate from a natural-gas transmission line, remove ethane and propane from it, and return the methane to the pipeline, as shown in Fig. 12-16.

## Cost data

|  | Cost per cubic meter |
|---|---|
| Feed | $0.06 |
| Price of $x_1$ returned to pipeline[†] | 0.0595 |
| $x_2$ | 0.08 |
| $x_3$ | 0.10 |
| Operating cost: |  |
| Extractor, per cubic meter of total flow | 0.001 |
| Separator, per cubic meter of pure propane | 0.003 |

[†]Since the methane has a lower heating value than the original feed, it is less valuable.

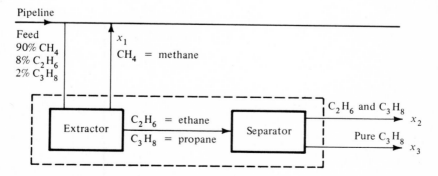

**Figure 12-16** Removal of propane and ethane from a natural-gas pipeline. The units of the flow rates, $x_1, x_2,$ and $x_3$, are in cubic meters per second.

*Restrictions*

The composition of the feed is 90% methane, 8% ethane, and 2% propane.
The maximum capacity of the extractor is 200 $m^3$/s of feed.
The maximum capacity of the separator is 3 $m^3$/s of pure propane.

Set up the objective function and constraints to maximize the profit of this plant.
**Ans.:** Constraints are

$$x_3 \leqslant 3$$
$$x_2 + x_3 \leqslant 20$$
$$x_1 - 9x_2 - 9x_3 = 0$$
$$x_2 - 4x_3 \geqslant 0$$

# REFERENCE

1. P. Vandenhoeck, "Cooling Hot Gases before Baghouse Filtration," *Chem. Eng.*, vol. 79, pp. 67–70, May 1, 1972.

# ADDITIONAL READINGS

Charnes, A., W. W. Cooper, and A. Henderson: *An Introduction to Linear Programming*, Wiley, New York 1953.

Dano, S.: *Linear Programming in Industry*, 3d ed., Springer, New York, 1965.

Dantzig, G. B.: *Linear Programming and Extension*, Princeton University Press, Princeton, N.J., 1963.

Garvin, W. W.: *Introduction to Linear Programming*, McGraw-Hill, New York, 1960.

Greenwald, D. U.: *Linear Programming*, Ronald Press, New York, 1957.

Hadley, G.: *Linear Programming*, Addison-Wesley, Reading, Mass., 1962.

Llewellyn, R. W.: *Linear Programming*, Holt, New York, 1964.

# COMPREHENSIVE PROBLEMS

This appendix presents some sample projects which apply principles studied in the text, e.g., economics, equation fitting, simulation, optimization, or a combination of them. These problems may be used as projects accompanying the study of the text material and running as a part-time effort all term. Many instructors devise their own similar problems based on their own engineering experiences. Some of the problems may carry over from one term to the next, with one team of students picking up the work where the preceding group left off.

Engineering students become proficient in solving short problems such as homework problems which require 45 min, but most professional engineering problems are long-term, requiring weeks or months for completion. It is therefore appropriate for senior-level or graduate-student engineers to gain some experience with comprehensive problems which require discipline to maintain progress over a longer period of time. Also, at the beginning of any long-term project there is the period of deliberation on how to start the problem—how to find the handle. Inexperienced engineers spend considerable time spinning their wheels and making false starts before focusing on a valid solution. Experience with comprehensive projects is the best means of developing proficiency in thought and work habits. Written or oral reports make good targets for completion and have their own benefit as well.

The projects that follow consist of a statement of the problem, which contains some or all of the required data. In certain cases property data will have to be extracted from handbooks. With some of the problem statements will be included some brief comments on the experiences of groups who have worked on the problem.

A-1. Optimum temperature distribution in a multistage flash-evaporation desalination plant

A-2. Heat recovery from exhaust air with an ethylene glycol runaround system

A-3. Design of a fire-water grid

A-4. Optimum thickness of insulation in a refrigerated warehouse

A-5. Simulation of a liquefied-natural-gas facility

A-6. Optimization of a natural-convection air-cooled condenser

A-7. Enhancing the heat-transfer coefficient of boiling in tubes

A-8. Recovery of heat from exhaust air using a heat pump

A-9. Simulation of a dehumidifier for industrial drying

A-10. Optimum gas pipeline when recovering power at terminus of the pipeline

A-11. Conserving energy by using a refrigerant mixture

A-12. Optimum air ejector for a steam-power-plant condenser

A-13. Optimizing a hot-oil loop in a petrochemical plant

A-14. Optimum heat pump for pasteurizing milk

## A-1 OPTIMUM TEMPERATURE DISTRIBUTION IN A MULTISTAGE FLASH-EVAPORATION DESALINATION PLANT

### Flash Desalination

One of the methods for water desalination is a distillation process using multi-stage flash evaporators, as shown schematically in Fig. A-1. Seawater flows first

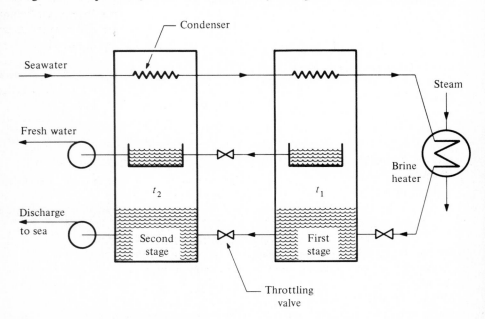

**Figure A-1** Multiflash desalination facility.

through heat exchangers, on which vapor condenses to form fresh water. After this preliminary heating of the seawater in the condensers, a steam heater elevates the temperature of the seawater to the maximum permitted by corrosion limitations. In passing through the first throttling valve some of the hot brine vaporizes, and this water vapor condenses on the tubes of the condenser and drops to the freshwater collection pan. From the first stage the brine and the fresh water flow through pressure-reducing valves into the next stage.

Figure A-1 shows a two-stage plant, but commercial plants have many more stages. A Foster-Wheeler plant in San Diego, for example, has nine stages.[1]

The temperature in a stage is defined as the condensing temperature of the vapor, which in turn is dictated by the amount of heat-transfer area in each stage of the condenser. The assignment is to specify the optimum number of stages and the heat-transfer area in each of the condensers for a minimum total cost over the 20-year life of the desalination plant, for a plant with a freshwater capacity of 50 kg/s.

*Design data*

Life, 20 years.

Interest on investment, 8 percent.

Heat-exchanger cost, $95 per square meter for both condensers and the brine heater.

$U$ value for both condensers and brine heater, 2800 W/(m² · K).

Seawater temperature, 15°C.

Steam temperature in brine heater, 150°C.

Maximum permitted temperature of brine leaving the brine heater, 120°C.

Steam for the brine heater costs 90 cents per gigajoule; each additional flashing stage costs $12,000.

Assume that the thermal properties of the brine are the same as those of pure water.

## Discussion

One approach is to start with a single-stage plant, which provides experience in setting up the thermodynamic and heat-transfer equations. An interesting observation is that the minimum stage temperature is the average of the seawater temperature and the maximum brine temperature, thus (15 + 120)/2 = 67.5°C. This minimum temperature occurs with a condenser of infinite area. The optimum condenser area results in a temperature that is a fraction of a degree higher than 67.5°C.

For a two-stage plant, infinite area in the two condensers results in stage temperatures that again divide the 15 to 120°C range equally, but this time into thirds. Thus the minimum temperature of the first stage is 85°C and in the second stage 50°C.

This problem can be extended into a two-level optimization. The first level is to determine the optimum temperatures in, say, a two-stage plant. The next level is to optimize the number of stages and for each different number of stages perform a new temperature optimization.

## A-2 HEAT RECOVERY FROM EXHAUST AIR WITH AN ETHYLENE GLYCOL RUNAROUND SYSTEM

### Introduction

Most large buildings have ventilating requirements which bring in outdoor air and exhaust an equal amount. When the outdoor temperature is low, the cost of heating the outdoor air before introducing it into the building is appreciable. This heating cost can be reduced by recovering some heat from the exhaust air. One method of recovering heat is to place a finned-coil heat exchanger in the exhaust airstream, another in the outdoor airstream and to pump a fluid between the coils, as shown in Fig. A-2. Water would be the first choice for a heat-transfer fluid, but to guard against freezeup at low outdoor temperature, an antifreeze such as ethylene glycol must be added.

If the system serves any purpose at all, it must save more money in heating

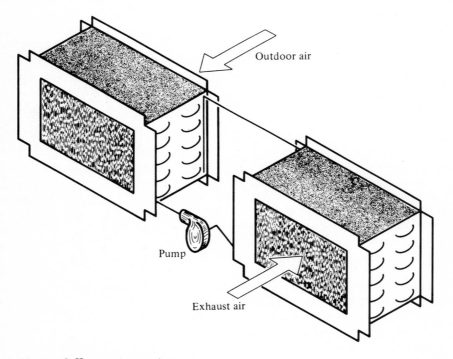

**Figure A-2** Heat-recovery system.

costs than it requires for its own amortization and operation. The first costs that are to be amortized include the costs of the coils, pump, piping, wiring, additional ductwork or revisions thereof, and additional cost of larger fans or motors, if needed. The operating costs include the power for the pump and additional fan power.

The potential savings should be evaluated for an optimum heat-recovery system. This optimum system consists of the most favorable combination of the following variables:

1. Length of the tubes in the coil
2. Height (or number of tubes high) of the coil
3. Number of rows of tubes deep (parallel to the path of the airflow)
4. Fin spacing
5. Glycol flow rate

The qualitative effects of these five variables are as follows. Increasing the length and number of tubes high increases the heat-transfer area but also the cost. Furthermore, increasing the length and number of tubes high increases the cross-sectional area for airflow, which reduces the velocity and decreases the air-side heat-transfer coefficient but also decreases the power required of the fan to force the air through the coil. Increasing the number of rows of tubes deep increases the heat-transfer area but increases the first cost of the coil as well as both the air and the glycol pumping cost. Spacing the fins closer together increases the cost of the coil and the air-pressure drop but also increases the heat-transfer area. Finally, a high flow rate of the ethylene glycol increases the glycol-side heat-transfer coefficient but also increases the pumping cost.

## Further Data and Assumptions

The size of the optimum system will pertain to a given airflow rate, and 3.0 m³/s has been chosen for both the outdoor-air and exhaust-air flow rate. The coil circuiting chosen is that of vertical headers feeding horizontal tube circuits in parallel, as shown in Fig. A-2. The flow of the glycol through the U bends is counter to the flow of air. Assume pure counterflow in the coils; thus the equations that represent this runaround loop are those used in Prob. 5-10.

*Further specifications*

The copper tubes have an OD of 16 mm and a wall thickness of 1 mm.
The tube spacing is 41 mm vertically and horizontally.
The average outdoor temperature is 5°C for 250 days of 24-h operation.
The life is 10 years, and the interest rate is 8 percent.
The electricity cost is 4 cents per kilowatthour.
The fan and pump motor efficiencies are 75 percent.

## The Optimization

The net saving to be maximized is the difference between the reduction in heating cost and the annual cost (amortized first cost plus operating cost) of the recovery system. The building is heated electrically, so as far as the ventilation air is concerned, electric-resistance heat must warm the incoming outdoor air to the building temperature of 24°C. The first cost to be amortized includes that of the coils, the pump and motor (assumed constant at $400), and the interconnecting piping (assumed constant at $150). No allowance for capital cost for the fan is provided because fans already exist in the system, and it is assumed that they would not have to be enlarged to overcome the additional pressure drop of the heat-recovery coils.

## Cost and Performance Equations

The equation for the first cost of each coil roughly reflects the proportionality of the cost to the mass of the coil,

$$\text{Cost} = 0.26[18 + (0.024)(NF + 500)W][L(NR + 1)] \quad \text{dollars}$$

where $NF$ = fin spacing along the tube, fins/m
$W$ = number of rows of tubes deep (in direction of airflow)
$L$ = tube length, m
$NR$ = number of layers of tubes high

The $U$ value of the coil based on outside (air-side) area is

$$\frac{1}{U_o} = \frac{0.0226}{V_a^{0.58}} + 0.0032 + \frac{0.0088NF + 0.185}{430V_{eg}^{0.8}}$$

where $U_o$ = $U$ value, W/(m$^2$ · K)
$V_a$ = face velocity of air through coil, m/s
$V_{eg}$ = velocity of ethylene glycol in tubes, m/s

The thermal resistance of the tubes and fins is 0.0032 (m$^2$ · K)/W.
Air pressure drop (DPA) in pascals is

$$\text{DPA} = (4.1\,W)(0.25 + 0.0016NF)(V_a^{1.75})$$

and the pressure drop of ethylene glycol (DPEG) in kilopascals is

$$\text{DPEG} = 5.2[0.15WL + 0.0875(W - 1) + 0.3](V_{eg}^{1.75})$$

In addition to the pressure drop of the ethylene glycol through the coil, some allowance should be made for the interconnecting piping, 15 kPa, for example.

## Variables

The five variables for which optimal values are sought and the ranges of their values are

| NF, fin spacing | 300, 400, 500, 600 fins/m |
| L, tube length | 0.4, 0.6, . . . , 3.4, 3.6 m |
| W, number of rows of tubes deep | 2, 4, 6, 8, 10 |
| NR, number of rows of tubes high | 4, 5, 6, . . . to $L/0.06$ |
| $V_{eg}$, velocity of ethylene glycol | 0.4, 0.8, 1.2, 1.6, 2.0 m/s |

## Assignment

Determine the combination of the above variables that results in the optimum economic solution.

## A-3 DESIGN OF A FIRE-WATER GRID

### Fire Control

Refineries and other chemical plants that process flammable substances must provide elaborate measures to prevent and fight fires.[2] Almost all such plants must be equipped with a fire-water distribution system that is generally constructed underground to supply hydrants throughout the plant. Typically, the plant is subdivided into various areas, remote from each other so that if a fire breaks out in one area it can be contained in that area. To serve the entire plant, the fire-water grid should be capable of providing a specified rate of water flow to any one area at a time.

Two challenges face the designer. For a given pressure at the outlet of the pump:

1. Select the pipe sizes so that the flow through the grid will supply the hydrants that surround a plant area with the required rate of water flow
2. Select the minimum first-cost combination of pipe sizes that meets the above requirement

Most designers of industrial fire-water grids are satisfied to achieve task 1, and even this assignment is a challenging one. Some designers use specially constructed electrical analogs in combination with a cut-and-try method of enlarging pipe sizes until each area individually can be blanketed with specified water flow. The analog must be a special one, because the fluid-flow conductor does not follow Ohm's law.

### Objective

The goal of this project is to select pipe sizes for a fire-water grid to achieve minimum first cost when the following conditions are specified:

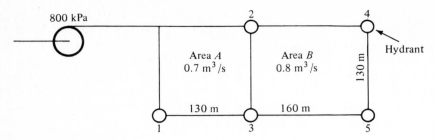

**Figure A-3** Fire-water grid.

1. Geometric layout of areas
2. Location of hydrants
3. Water pressure at the pump outlet
4. Minimum flow rate required by each operating area

Although in actual plants there may be a dozen operating areas with hydrants distributed along the pipes, a suggested grid for this type of problem is shown in Fig. A-3.

## Operating and Cost Data

The supply pressure of the pump is 800 kPa. The grid must supply a minimum of 0.7 m$^3$/s to area $A$ and 0.8 m$^3$/s to area $B$, but not simultaneously. Area $A$ is served by hydrants 1, 2, and 3, while area $B$ is served by hydrants 2, 3, 4, and 5. The flow characteristics of the hydrants are

$$\text{Flow rate, m}^3\text{/s} = 0.01 \sqrt{\text{pressure, kPa}}$$

The installed cost of the pipe is 30 cents per meter of length for each millimeter of diameter. The trenching cost is $80 per meter, but since it is independent of the pipe size, it is a constant and does not affect the optimum pipe-size selection. Neglect operating cost, because one hopes the number of hours of pump operation will be negligible.

## Discussion

Finding the optimum grid design reduces to two separate optimizations, optimizing one area at a time. The adequate system that serves area $A$ alone will probably be inadequate for area $B$ alone, so in this case the optimum design that serves area $B$ should be checked to see whether it is adequate for area $A$.

The problem reduces to one involving equality constraints and thus is soluble by the method of Lagrange multipliers, although estimating the trial values of the $\lambda$'s in the set of nonlinear equations is difficult.

The optimization is likely to shrink the size of some legs of the grid to zero, which may be undesirable from the security standpoint. For this reason another set of constraints is warranted which specifies the minimum diameter of all pipe sections to be 150 mm, for example.

Another refinement that adds more constraints is that there should be a minimum pressure at the hydrant, perhaps 400 kPa. The reason for this minimum is that adequate pressure must be available to send a stream of water at least into the center of the area, and perhaps farther, in order to provide some overlap of water coverage.

## A-4 OPTIMUM THICKNESS OF INSULATION IN REFRIGERATED WAREHOUSES

### Optimum Insulation Thickness

Deciding upon the optimum insulation thickness is a classic engineering problem. When heat transfer into or out of a controlled-temperature space results in cooling or heating cost, application of insulation will reduce these costs. Each additional unit thickness of insulation, however, is progressively less effective, and the point is reached where additional thickness of insulation is not compensated for by the reduction in operating cost.

A large food company suggested that some of the standard thicknesses of insulation used in refrigerated warehouses should be restudied. During recent years the cost of energy has increased at a more rapid rate than the cost of insulation and the first cost of the refrigeration plant. Thicker insulation than former standards may now be justified.

### Cost and Engineering Data

The optimization will be performed for the insulation thickness in the roof, for which there is considerable latitude. (The walls are now usually constructed of prefabricated panels, where there is limited choice of insulation thickness.) The warehouse temperature is $-18°C$, and the average outdoor temperature during the year will be between 5 and $10°C$, depending upon the climate.

The insulation board is available in 50-, 75-, 100-, and 125-mm thicknesses. To develop thicknesses greater than 125 mm, one or two additional layers of insulating board are required.

#### Additional data

Cost of insulation, 5 cents per square meter of area for each millimeter of thickness

Installation cost, first layer, $2.50 per square meter; second and third layers $1.50 per square meter

First cost of refrigeration equipment, $600 per kilowatt of refrigeration capacity

Power required by refrigeration plant, 0.6 kW of electric power per kilowatt of refrigeration

Cost of electric energy (depends upon local rates), usually between 3 and 6 cents per kilowatthour

Conductivity of insulation $k$, 0.04 W/(m · K)

The expected life of the plant is 15 years; the rate of interest is 9 percent; and for tax purposes the facility may be written off in 10 years (suggest straight-line depreciation). Federal income tax is 50 percent.

Determine the optimum thickness.

## Discussion

To determine the influence of the power cost and the average outdoor temperature on the optimum thickness, it is suggested that the optimization be performed for several values in the complete range of outdoor temperatures and power costs. The cost associated with the application of an additional layer of insulation is likely to inhibit the optimal thickness from exceeding 250 mm, or two full-size layers of insulation.

## A-5 SIMULATION OF A LIQUEFIED NATURAL GAS FACILITY

### Need for Liquefying Natural Gas

The principal reasons for liquefying natural gas are for shipping and for storage. When natural gas is shipped across an ocean, it is economical to convert it into liquid form since 600 times more gas can be contained in a given volume than in the gaseous form at standard atmospheric pressure. To hold liquefied natural gas (LNG) at atmospheric pressure, the liquid temperature must be approximately −155°C.

The need for storage of natural gas often arises in the following manner. Gas utility companies are distribution companies, and they purchase natural gas from transmission companies. The transmission companies assess demand charges, so it is often to the advantage of the distribution company to buy gas from the transmission line during periods of low demand and store the gas as LNG for use during the peak winter periods.

### The LNG Plant

Figure A-4 shows a schematic diagram of a simplified LNG plant. In the cycle of Fig. A-4, gas from the transmission line enters the liquefaction plant at 1000

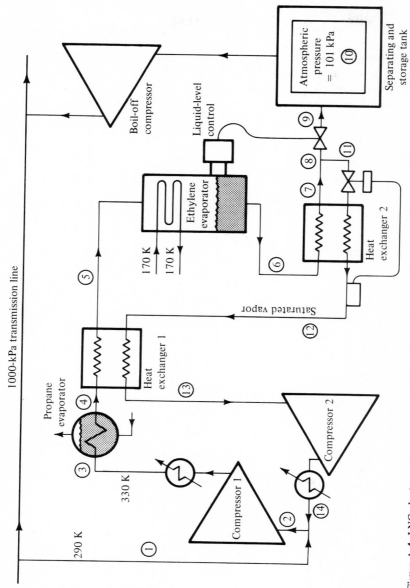

**Figure A-4** LNG plant.

kPa and 290 K. The stream at point 1 joins with recycled gas from compressor 2 and is pumped up to the high pressure by compressor 1. The gas is then de-superheated in the propane evaporator and heat exchanger 1. Liquefaction takes place in the ethylene evaporator but at a high pressure. The saturated liquid at point 6 is subcooled in heat exchanger 2 by transferring heat to low-pressure fluid boiling in the other stream of the heat exchanger. The subcooled liquid at point 8 passes through a throttling valve and expands to atmospheric pressure. Some of this liquid flashes into vapor, and this vapor is recompressed by the boil-off compressor back to the transmission-line pressure.

Some pressures and temperatures are noted in Fig. A-4, and additional property equations for the gas (which is assumed to be methane) and performance characteristics of the components are provided below.

*Properties*

Enthalpy of saturated liquid

$$h_f, \text{ kJ/kg} = -219.6 + 7.025T - 0.034965T^2 + 0.00011321T^3$$

Enthalpy of saturated vapor

$$h_g, \text{ kJ/kg} = 790.13 - 4.0137T + 0.058268T^2 - 0.00019465T^3$$

Specific heat of superheated vapor

$$c_p, \text{ kJ/(kg} \cdot \text{K)} = 2.114 + 0.000188p - (1.76 \times 10^{-8})p^2$$

Saturation pressure

$$\ln p = -\frac{1026}{T} + 13.8$$

where $p$ = pressure, kPa
$\qquad T$ = temperature, K

*Component performance*

$$UA = \begin{cases} 2.64 \text{ kW/K} & \text{propane evaporator} \\ 4.22 \text{ kW/K} & \text{heat exchanger 1} \\ 5.275 \text{ kW/K} & \text{heat exchanger 2} \\ 42.2 \text{ kW/K} & \text{ethylene evaporator} \end{cases}$$

In the ethylene evaporator assume that the temperature of methane is at the saturation temperature throughout when determining the mean-temperature difference.

Compressor 1: $\qquad \dfrac{p_{\text{disch}}}{p_{\text{inlet}}} = 7.5 - 150.5 \dfrac{(w, \text{kg/s}) \sqrt{T_{\text{inlet}}, \text{K}}}{p_{\text{inlet}}, \text{kPa}}$

Compressor 2: $\qquad \dfrac{p_{\text{disch}}}{p_{\text{inlet}}} = 5.6 - 142.8 \dfrac{(w, \text{kg/s}) \sqrt{T_{\text{inlet}}, \text{K}}}{p_{\text{inlet}}, \text{kPa}}$

## Discussion

Using the system-simulation procedures of Chap. 6 with the aid of a computer program, a team found several key results.

Liquefaction rate, 0.64 kg/s
High pressure, 3610 kPa
Pressure entering compressor, 2400 kPa

## A-6 OPTIMIZATION OF A NATURAL-CONVECTION AIR-COOLED CONDENSER

Air-cooled condensers that use propellor fans to blow air over finned heat exchangers are commonly used in process industries, such as refineries. The electric motors, fans, bearings, and gears or belts, if they are used, require maintenance and sometimes fail if not properly maintained. Furthermore the fan motors require energy, and natural-convection air-cooled condensers are now often used.[3] Figure A-5 shows a sketch of such a condenser, and Fig. A-6 is a sketch of the finned coil.

## Objective

Design for minimum first cost an air-cooled condenser that rejects 140 kW of heat. Specifically select the height $h$, the dimensions of the square coil and stack $b$, and the number of rows of tubes high in the heat-exchanger section.

*Performance data*

Temperature of ambient air, 35°C.
Temperature of condensing fluid, 90°C.
Choices of number of rows of tubes high, 1 to 6.
Choices of $b$ dimension, 1.4 to 3.0 m in 0.2-m increments.
Heat-transfer area on condensing-fluid side of heat exchanger per square meter of face area per row of tubes high, 1.3 m².
Effective heat-transfer area on the air side of the heat exchanger per square meter of face area per row of tubes high, 12.5 m².[†]
Neglect the resistance to heat transfer through the tube.

[†]The actual surface area is greater than indicated, but the effective area accounts for the fin efficiency.

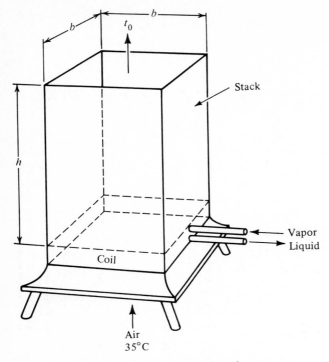

**Figure A-5** Natural-convection air-cooled condenser.

Heat-transfer coefficient of condensing fluid, 2.0 kW/(m² · K).

Heat-transfer coefficient on air side, $0.042V^{0.4}$ kW/(m² · K), where $V$ is the face velocity of air in meters per second.

Pressure drop through the coil, $\Delta p$, Pa = $2.2V^{1.8}(1.5 + 0.7n)$, where $n$ is the number of rows of tubes.

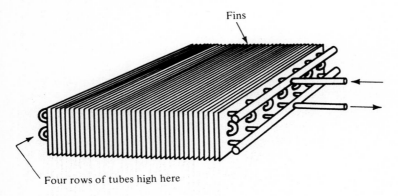

**Figure A-6** Condenser coil.

Pressure drop due to friction of the air flowing through the stack must be considered; suggest a friction factor of 0.02.

*Costs*

Stack, including cost of supports, $45 per square meter of area
Coil, $(190) (b^{3.2}) (n^{0.6})$ dollars

## Discussion

An increase in the height $h$ improves the stack effect and thus increases the airflow rate but adds to the cost. An increase in the coil and stack dimension $b$ reduces the pressure drop through the coil and stack but increases the cost. An increase in the number of rows of tubes increases the heat-transfer area and outlet temperature of air $t_o$ but increases the pressure drop. An increase in air velocity over the coil increases the heat-transfer coefficient on the air side but reduces $t_o$.

Not all combinations of $b$ and $n$ may be workable.

The stack effect due to the difference in densities of the ambient air and the air in the stack is responsible for airflow and compensates for the air-pressure drop through the coil and in the stack itself.

## A-7 ENHANCING THE HEAT-TRANSFER COEFFICIENT OF BOILING IN TUBES

The heat-transfer coefficient of a boiling fluid inside a tube varies throughout the length of the tube as a function of the quality (fraction of vapor). A typical profile is shown in Fig. A-7. In order to improve the overall performance of the heat exchanger a modified evaporator, as shown in Fig. A-8, extracts vapor at two positions and injects it elsewhere along the tube in order to maintain the quality near 0.7 or 0.8 and thus take advantage of the high heat-transfer coefficient that results.

## Objective

Determine the positions of vapor extraction and the flow rates of vapor that result in the optimum mean heat-transfer coefficient if the local coefficients are as shown in Fig. A-8. Specifically, determine $w_1/w$, $w_2/w$, $y/L$, and $z/L$ for the maximum mean coefficient. Assume that the vapor and liquid move at the same velocity.

## Discussion

The analysis is complicated by the fact that the heat flux changes along the tube because the coefficient is not constant. Furthermore, the heat flux is influenced

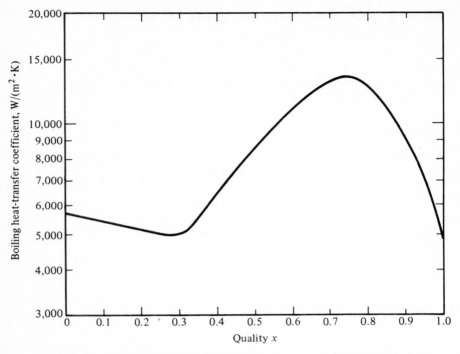

**Figure A-7** Boiling heat-transfer coefficient inside a tube. (*From J. M. Chawla, "A Refriger-ation System with Auxiliary Liquid and Vapour Circuits," Int. Inst. Refrig., Meet Comm. II and III, London, 1970; used by permission.*)

by the local temperature of the fluid being cooled. To simplify, assume that the rate of evaporation is uniform along the length; thus the quality varies linearly with distance from 0 to 1.0.

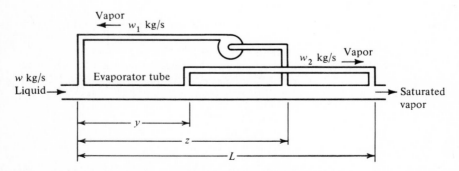

**Figure A-8** Evaporator with enhanced heat-transfer coefficients.

# A-8 RECOVERY OF HEAT FROM EXHAUST AIR USING A HEAT PUMP

In the interest of energy conservation, heat exchangers (rotary-wheel and run-around systems, as in Prob. A-2) are sometimes installed in buildings to recover some of the heat of exhaust air in winter and precool outdoor ventilation air in the summer. These heat exchangers function because the exhaust air is at a higher temperature than the incoming outdoor ventilation air in winter and at a lower temperature than the incoming air in summer.

To intensify the heat-transfer process, the installation of a heat pump is proposed as shown in Fig. A-9. The building is electrically heated, so that if the heat pump does not bring the temperature of the incoming air up to 35°C (the supply air temperature), electric-resistance heat is used to supplement the heat pump.

## Objective

Determine the optimum heat pump, namely, the size of compressor, condenser, and evaporator that provides a minimum total present worth of costs (first cost plus present worth of power costs).

## Data

$U$ values of condenser and evaporator coils, 25 W/(m² · K) based on air-side area
Cost of coils, $50 per square meter of air-side area
Compressor cost, including motor, $120 per motor kilowatt
Power cost, 3 cents per kilowatthour
Interest rate, 10 percent
Economic life, 10 years

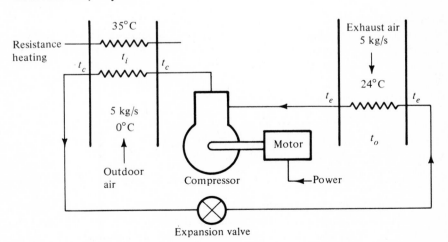

**Figure A-9** Heat pump for recovering heat from exhaust air and preheating incoming ventilation air.

Number of hours of operation per year, 4000
Average outdoor temperature during the 4000 h, 0°C
Air flow rate through both coils, 5 kg/s

The performance characteristics of the compressor can be expressed as a coefficient of performance (COP), where

$$COP = \frac{\text{refrigeration rate, kW}}{\text{electric power to compressor motor, kW}}$$

The COP is a function of the evaporating and condensing temperatures, $t_e$°C, and $t_c$°C, and can be represented by

$$COP = 7.24 + 0.352t_e - 0.096t_c - 0.0055t_e t_c$$

## Assignment

Determine the optimal combination of evaporator area, condenser area, and compressor motor input power that results in the minimum total present worth of costs. Also determine $t_e$, $t_c$, $t_i$, and $t_o$ at this optimum.

## Discussion

If the heat pump can be justified for heating operation alone, it has the possibility of providing additional savings by operating during the summer.

The temperature of the supply air entering the building is specified as 35°C, but during some periods of the heating season it may be desirable for this temperature to be above or below 35°C. This condition would revise the assignment somewhat.

The optimization is conducted on the basis of a constant outdoor air temperature of 0°C. During actual operation the COP of the system will be higher than average when the outdoor temperature is above 0°C and less heat recovery is needed. Conversely, when the outdoor temperature is below 0°C, the COP drops and more heat-recovery capacity can be used.

## A-9 SIMULATION OF A DEHUMIDIFIER USED FOR INDUSTRIAL DRYING

A traditional method of drying many materials (food products, grain, paper, etc.) has been to heat ambient air and pass it over the material to be dried. Some industries have been forced to shift from heat obtained from natural gas and oil to the use of electricity. Instead of using the electricity in a resistance heater, some applications have employed a heat pump, as shown in Fig. A-10. The evaporator of the heat pump cools and dehumidifies the air and then reheats the

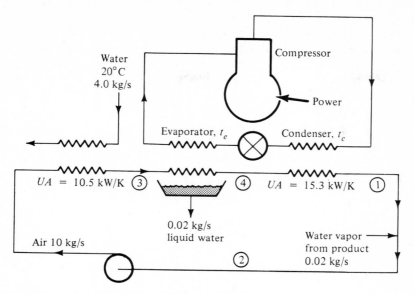

**Figure A-10** Heat-pump dehumidifier for industrial drying.

air with the refrigerant condenser. Next the air passes over the material to be dried. Energy is added to the air by the heat pump and also by virtue of the water entering the airstream as vapor and leaving as liquid. A water-cooled heat exchanger removes this energy.

*Data*

Airflow rate, 10 kg/s
Rate of addition of water vapor, 0.02 kg/s
Temperature of water entering water-cooled heat exchanger, 20°C
Flow rate of water entering heat exchanger, 4.0 kg/s
$UA$ of water-cooled heat exchanger, 10.5 kW/K
$UA$ of condenser, 15.3 kW/K

**Compressor performance**  The pumping capacity of the compressor expressed in terms of heat-transfer rate at the evaporator is

$$q_e = 126.9 + 7.683t_e - 1.134t_c - 0.05516t_e t_c \quad \text{kW}$$

and the power requirement of the compressor is

$$P = 15.05 - 1.266t_e + 0.283t_c + 0.0299t_e t_c \quad \text{kW}$$

where $t_e$ = evaporating temperature, °C
$t_c$ = condensing temperature, °C

**Evaporator** The evaporator is a heat exchanger where both sensible heat and mass are transferred from the air to the surface of the heat exchanger, which is wet because of the dehumidification. The heat then flows through the metal and thence to the refrigerant. Data on the evaporator are

Refrigerant-side area $A_r$, 15 m$^2$
Refrigerant-side heat-transfer coefficient $h_r$, 2.2 kW/(m$^2$ · K)
Neglect the heat-transfer resistance of the metal
Air-side area $A_a$, 120 m$^2$
Air-side convection heat-transfer coefficient, $h_a$ = 0.09 kW/(m$^2$ · K)

**Heat and mass transfer in the evaporator** The processes occurring in the evaporator coil can be visualized as shown in Fig. A-11. The driving force for the transfer of sensible heat is the difference in dry-bulb temperatures of the air and the wetted surface. If the arithmetic-mean difference is assumed to be sufficiently accurate,

$$q_s = h_a A_a \left( \frac{t_3 + t_4}{2} - \frac{t_{ws3} + t_{ws4}}{2} \right)$$

where $q_s$ is the rate of sensible-heat transfer in kilowatts. The rate of latent heat transfer[4] is

$$q_L = \frac{h_a}{c_p} A_a \left( \frac{W_3 + W_4}{2} - \frac{W_{ws3} + W_{ws4}}{2} \right) h_{fg} \quad \text{kW}$$

where $W_3, W_4$ = humidity ratio of air, (kg water)/(kg dry air)
$W_{ws3}, W_{ws4}$ = humidity ratio of saturated air at wetted-surface temperature, (kg water)/(kg of dry air)
$c_p$ = specific heat of air, kJ/(kg · K)
$h_{fg}$ = latent heat of water (suggest using a constant value of 2450 kJ/kg)

On the refrigerant side this combination of $q_L$ and $q_s$ equals the rate of heat absorbed by the refrigerant $q_e$, and

$$q_e = h_r A_r \left( \frac{t_{ws3} + t_{ws4}}{2} - t_e \right)$$

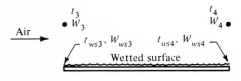

**Figure A-11** Heat and mass transfer at the evaporator coil.

The temperature of the wetted surface, $t_{ws3}$ for example, must adjust itself so that the combined rate flow of sensible plus latent heat equals the rate flow of heat to the refrigerant at that position.

## Assignment

Simulate this system in the sense of computing all pertinent operating variables, including the conditions of the air throughout the cycle; heat-transfer rates at the evaporator, condenser, and water-cooled heat exchanger; and the power required by the compressor.

## Discussion

The specification of the rate of moisture removal (0.02 kg/s) permits a simplification in this simulation in that the humidity ratios of the air throughout the cycle can be left free to float until the remainder of the variables are simulated. An extension of this problem is to express the rate of moisture removal as a function of the mass transfer of water from the product to the air. The humidity ratios of the air must then adjust to provide a water balance.

## A-10 OPTIMUM GAS PIPELINE AND PUMPING FACILITY WHERE WORK CAN BE RECOVERED AT DESTINATION

In a natural-gas (methane) pipeline 120 km long (Fig. A-12) there is a need for power at the receiving point, so the system is being designed with a gas turbine at the receiving end to expand the gas from the pipeline back down to 100 kPa.[5] Compression to a high pressure permits a high-pressure drop through the pipeline and also results in dense gas; both reduce the pipe size. On the other hand, there is additional energy required for compression, and not all this energy is re-

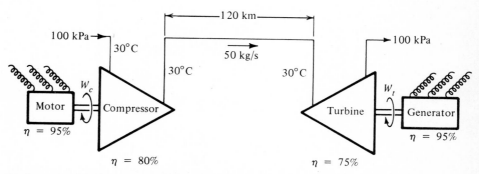

**Figure A-12** Natural-gas pipeline where work is recovered at terminal position.

covered at the turbine due to inefficiencies of the electric motor, compressor, and turbine, as well as pipeline losses.

## Data

Methane enters the compressor at 100 kPa.
Methane leaves the turbine at 100 kPa.
All the power generated by the turbine-generator can be used.
Gas enters the compressor at 30°C, is cooled after the compression, and remains
    at 30°C throughout the pipeline.
Methane flow rate is 50 kg/s.
The efficiencies of the electric motor and generator are 95 percent.
The efficiency (with respect to the isentropic process) of the compressor is
    80 percent and of the turbine is 75 percent.

## Costs

Electric motor and generator first cost, $50 per kilowatt output
Compressor first cost, $125 per kilowatt input (input designated $W_c$)
Value of electricity at compressor end of pipeline, 3 cents per kilowatthour
Value of electricity at turbine end of pipeline, 4 cents per kilowatthour
Turbine first cost, $150 per kilowatthour output (output designated $W_t$)
Pipe cost in dollars per meter length, $300D^{1.6}$, where $D$ is the pipe diameter
    in meters
Life of facility, 15 years
Interest rate, 10 percent
Operation, continuous

**Pipeline pressure drop** Because the pressure changes as the gas flows through the pipeline, the density and velocity also change. For an element of length $dL$, the pressure drop $dp$ is expressed by

$$dp = \frac{f \, dL}{D} \frac{V^2}{2} \rho$$

where $p$ = pressure, Pa
    $f$ = friction factor, dimensionless (suggest $f = 0.02$)
    $L$ = length, m
    $D$ = diameter, m
    $V$ = velocity, m/s
    $\rho$ = density, kg/m$^3$

## Assignment

Determine the optimum combination of sizes of compressor, turbine, and pipe to achieve the minimum sum of the first costs and present worth of operating costs.

# A-11 CONSERVING ENERGY BY USING A REFRIGERANT MIXTURE

When a single refrigerant condenses and evaporates at a constant pressure between saturated liquid and saturated vapor, the temperature remains constant. A binary solution, on the other hand, experiences a change in temperature, as pointed out in Chap. 5. When a refrigeration system cools a fluid stream through a large temperature range or the system rejects heat to a fluid that increases in temperature as it passes through the condenser, it may be possible to conserve compression power by using a binary solution.[6] Some applications of this concept have been made in the cryogenic and petrochemical industries.[7]

The refrigeration cycle adaptable to refrigerant mixtures employs the usual components (compressor, condenser, evaporator, and expansion device) and a heat exchanger as well, as shown in Fig. A-13. A mixture of refrigerants R-12 and R-114 will be explored.

## Data

$$UA = \begin{cases} 5 \text{ kW/K} & \text{evaporator} \\ 5 \text{ kW/K} & \text{condenser} \\ 0.3 \text{ kW/K} & \text{heat exchanger} \end{cases}$$

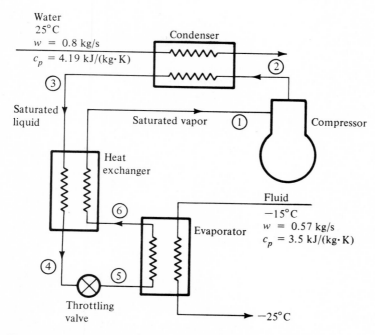

**Figure A-13** Refrigeration system using a mixture of refrigerants.

The condenser iş water-cooled; water enters at 25°C with a flow rate of 0.8 kg/s.
The evaporator fluid is cooled from −15 to −25°C. Its flow rate is 0.57 kg/s, and
  its specific heat is 3.5 kJ/(kg · K).
The refrigerant is saturated liquid at point 3 and saturated vapor at point 1.
The compressor has adjustable capacity which is regulated to provide the speci-
  fied refrigeration rate in the evaporator.

For saturated pressure

R-12:    $\ln p = 14.861 - 2498.3/T$

R-114:    $\ln p = 15.407 - 2993.2/T$

where $p$ = pressure, kPa
$\quad T$ = temperature, K

For enthalpy of saturated liquid

$$h_f = \begin{cases} 200 + 0.925t + 0.00081t^2 & \text{R-12} \\ 200 + 0.9545t + 0.00116t^2 & \text{R-114} \end{cases}$$

where $h_f$ = enthalpy of liquid, kJ/kg
$\quad t$ = temperature, °C

For enthalpy of saturated vapor

$$h_g = \begin{cases} 351.5 + 0.4283t - 0.00071t^2 & \text{R-12} \\ 337.4 + 0.6234t - 0.000086t^2 & \text{R-114} \end{cases}$$

where $h_g$ = enthalpy of saturated vapor, kJ/kg
$\quad t$ = temperature, °C

Work of compression

$$\Delta h = \begin{cases} 188\left(1 - \dfrac{14.861 - \ln p_2}{14.861 - \ln p_1}\right) & \text{R-12} \\[2mm] 158\left(1 - \dfrac{15.407 - \ln p_2}{15.407 - \ln p_1}\right) & \text{R-114} \end{cases}$$

where $\Delta h$ = work of compression, kJ/kg
$\quad p_2$ = discharge pressure (total pressure), kPa
$\quad p_1$ = suction pressure (total pressure), kPa

When compressing a mixture, the work of compression is found by propor-
tioning according to the mass fraction of the constituents in the mixture. The
mixture is assumed to be ideal, and Dalton's and Raoult's laws apply. The molecular weights of R-12 and R-114 are 120.93 and 170.94, respectively.

## Assignment

For this specified refrigeration duty (approximately 20 kW), determine the composition of the mixture that results in minimum power requirements at the compressor.

## Discussion

The optimum mixture is in the neighborhood of 50 percent of each. The heat exchanger causes an appreciable improvement in the performance of a mixed-refrigerant cycle but is of little benefit in a single-refrigerant system. After determining the optimum mixture, a further optimization would be possible in which the distribution of heat-transfer areas between the condenser, evaporator, and heat exchanger could be performed.

## A-12  OPTIMUM AIR EJECTOR FOR A STEAM-POWER-PLANT CONDENSER

The pressure of steam at the exhaust of the turbine and in the condenser of a steam power plant is below atmospheric, so air seeps into the system through any unavoidable leaks. The presence of air and other noncondensables results in a higher exhaust pressure from the turbine than would otherwise prevail, reducing both the power delivered by the turbine and the cycle efficiency. Provisions are made to remove air continuously by the use of such devices as an air-ejector system, shown in Fig. A-14. The function of the ejection system is to extract a sample of air and water vapor out of the main condenser and ultimately to reject the air in the mixture to the atmosphere. The removal of air is to be accomplished with a low loss of steam.

In the system shown in Fig. A-14, 0.12 kg/s of air-vapor mixture that contains 5 percent air by mass is extracted from the main condenser, and the mixture that is vented to atmosphere should contain 50 percent air, by mass. The main condenser operates with a total pressure of 6.6 kPa, and the aftercondenser operates at 101 kPa. Cooling water with a flow rate of 7 kg/s enters the intercondenser at a temperature of 35°C and passes in series through the intercondenser and then the aftercondenser.

## Performance of condensers

The performance of the intercondenser and the aftercondenser can be approximated by assuming that they are counterflow heat exchangers, where the temperature on the condensing side is that of the saturation temperature of water at the partial pressure of the water vapor. Thus, if the pressure at one point in the

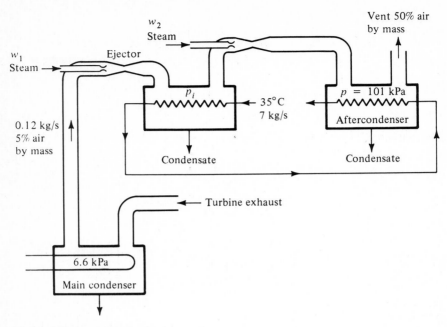

**Figure A-14** Two-stage air ejector.

intercondenser is 20 kPa and the partial pressure of the air is 2 kPa, the condensing pressure will be the saturation pressure at 18 kPa or 57.8°C. Use the log-mean temperature difference in the condenser with an applicable $U$ value of 1.4 kW/(m² · $K$).

## Assignment

Select the heat-transfer areas of the intercondenser and aftercondenser, the flow rates of motive steam $w_1$ and $w_2$, and the intermediate pressure $p_i$ so that the total present worth of all costs is a minimum.

## Further Data

Condenser cost, $150 per square meter of heat-transfer area
Cost of motive steam, 60 cents per megagram
Interest rate, 10 percent.
Life of facility for economic evaluation, 6 years.
Hours of operation per year, 4000.
The ratios of motive steam flow to pumped flow for the ejectors are given by

Low stage:  $\qquad$ Ratio (kg/s)/(kg/s) $= 0.19 \left(\dfrac{p_i}{6.6}\right)^{2.2}$

High stage:
$$\text{Ratio (kg/s)/(kg/s)} = 0.14 \left( \frac{101}{p_i} \right)^{2.2}$$

The combined first cost of the ejectors is approximately constant in the expected range of sizes to be examined.
The results of Prob. 4-16 may be useful.

## Discussion

The optimal design is likely to be one where more heat-transfer area is placed in the intercondenser than in the aftercondenser, since whatever steam (both from the main condenser and from the low-stage ejector) can be condensed at the intercondenser will not have to be pumped by the high-stage ejector. Often in multistage compressions the intermediate pressure is the geometric mean of the suction and discharge pressures. The economics of this system will shift the optimum intermediate pressure to a value lower than the geometric mean.

## A-13 OPTIMIZING A HOT-OIL LOOP IN A PETROCHEMICAL PLANT

One method of distributing heat in refineries and petrochemical plants is to heat the oil in a central furnace and pump it to various heat exchangers where heating is required. Figure A-15 shows a hot-oil loop that serves three reboilers. The furnace, reboilers, and pump are to be selected so that the total present worth of costs for the economic life of the facility is a minimum.

*Oil data*

Oil chemically stable up to 370°C
Specific heat, 2.1 kJ/(kg · K)
Density, 850 kg/m³

*Reboiler data*

|  | Duty, kW | Temperature of fluid being heated, °C | U value, W/(m² · K) |
|---|---|---|---|
| Reboiler 1 | 1800 | 200 | 450 |
| 2 | 1600 | 150 | 450 |
| 3 | 2400 | 175 | 450 |

Temperature of fluid being boiled in a given reboiler is essentially constant.
Cost of the reboilers is $90 per square meter of heat-exchanger area.

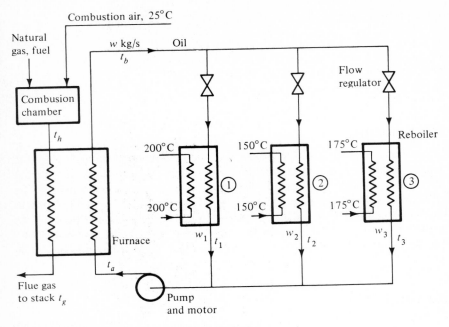

**Figure A-15** Hot-oil loop in a petrochemical plant.

*Furnace data*

Treat the furnace as a combination of a combustion chamber and a counterflow
heat exchanger.

Combustion air is heated to temperature $t_h$ in the combustion chamber and
then flows over heat-exchanger surface.

$U$ value of furnace heat exchanger, 30 W/(m²· K).

First cost of furnace, $30 per square meter of heat-exchanger area.

Cost of heating by natural gas, $1.20 per gigajoule.

Flow rate of combustion gas, 6.0 kg/s.

Specific heat of combustion gas, 1.1 kJ/(kg · K).

*Pump, motor, and electric data*

Efficiency of pump, 85 percent

Efficiency of electric motor, 90 percent

First cost of pump, $160 for each kilogram per second of oil flow

First cost of electric motor, $50 per kilowatt shaft-power output

Cost of electricity, 3.2 cents per kilowatthour

The pressure buildup required of the pump is a function of the specific design of the furnace, reboilers, and piping system, but as an approximate measure

of the pressure drop throughout the loop, use

$$\Delta p, \text{ kPa} = 8.2w^{1.8}$$

where $w$ is the system flow rate in kilograms per second.

*Economic data*
Plan on full-time operation for the 5-year economic life of the facility.
Interest rate, 10 percent

## Assignment

Optimize the system so that the minimum total present worth of costs results
for the economic life of the facility. Specifically, determine the optimum values
of the total oil flow rate and the flow rate in each reboiler; the leaving temperature of oil from each reboiler; $t_a$, $t_b$, $t_h$, $t_g$; the system oil-pressure drop; and
the size of the electric motor.

## Discussion

Since the dominant cost is that of the gas fuel, the optimization will drive
toward reduction of this cost. In particular the temperatures of the oil leaving
the reboilers will drop very close to the reboiler temperatures in an effort to reduce $t_a$, which in turn permits a low $t_g$ and thus a reduced loss of energy in the
leaving flue gases. A further extension that designers of these systems would
consider is a regenerative heat exchanger to recover more heat from the flue
gases and thus reduce the fuel cost.

## A-14  OPTIMUM HEAT PUMP FOR PASTEURIZING MILK

The essential requirement in the pasteurization process of milk is to bring the
temperature up to 73°C and hold it for approximately 20 s. The milk arrives at
the dairy from the tank truck at a temperature of 7°C and is to be delivered
from the pasteurizing plant to the packaging operation at a temperature of 4°C.

The traditional pasteurizing cycle uses a steam or hot-water heater to bring
the temperature of the milk to 73°C; then the milk flows through a water-cooled
heat exchanger and a refrigerant evaporator. The heat from the refrigeration
plant is rejected to the atmosphere by a cooling tower or an air-cooled condenser.

In the interest of conserving energy, the possibility of using a heat pump is
now sometimes considered.[8] One possible cycle is shown in Fig. A-16. The incoming milk flows first through a regenerative heat exchanger and then to the
heater, which is also the forecondenser of the heat pump. This forecondenser
elevates the temperature of the milk to 73°C. Thereafter the milk flows through

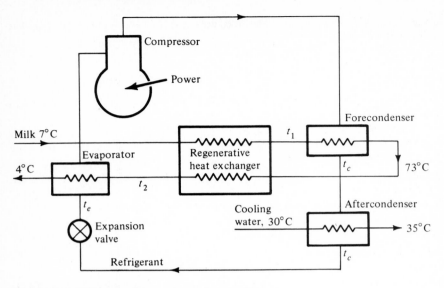

**Figure A-16** Milk pasteurizing system using a heat pump.

the other side of the regenerative heat exchanger and finally through the evaporator of the heat pump.

*Data*

Flow rate of milk, 4 kg/s for 4 h/day

Specific heat of milk, 3.75 kJ/(kg · K)

Cost of all heat-exchanger surfaces (evaporator, condensers, and regenerative heat exchanger), $95 per square meter

$U$ value of regenerative heat exchanger, 500 W/(m² · K)

$U$ values of refrigerant evaporator, forecondenser, and aftercondenser, 600 W/(m² · K)

Coefficient of performance of heat pump (ratio of the evaporator capacity to the compressor power), 75 percent of the Carnot COP for prevailing evaporating and condensing temperatures

First cost of the compressor and motor, $120 per kilowatt of power

Cost of electricity, 3.5 cents per kilowatthour

Interest rate, 9 percent

Economic life of the plant, 6 years

Cooling water enters the aftercondenser at 30°C and leaves at 35°C; flow rate of water is set to remove desired rate of heat flow.

## Assignment

Optimize the system to provide the minimum total present worth of costs for the economic life of the plant. Specifically, determine the areas of all heat exchangers, the power required by the compressor, and temperatures $t_1$ and $t_2$ that result in the optimum.

## Discussion

The optimum should occur when most of the heat exchange is accomplished in the regenerator.

The overall energy balance of the cycle indicates that because energy flows into the system through the use of compression work and because the milk leaves at a lower temperature than it enters, there must be a corresponding rejection of energy elsewhere. The heat rejection takes place at the aftercondenser. An alternate cycle to explore is one where the heat is rejected by a water-cooled heat exchanger in the milk stream rather than in the refrigerant stream.

## REFERENCES

1. W. F. Gardner, "Desalination Test Module," *Heat Eng.*, Foster-Wheeler, Corp., March-April, 1967 and September-October 1968.
2. A. M. Woodard, "How to Design a Plant Firewater System," *Hydrocarbon Process.*, October 1973, pp. 103–106.
3. P. T. Doyle and G. J. Benkly, "Use Fanless Air Coolers," *Hydrocarbon Process.*, July 1973, pp. 81–86.
4. *ASHRE Handbook, Fundamentals Volume*, chap. 3, American Society of Heating, Refrigerating, and Air-Conditioning Engineers, New York, 1977.
5. G. E. Zarnicki, L. A. Repin, and V. A. Elema, "Refrigeration by Utilizing the Pressure of Natural Gas Pumped through Pipelines," *Cholod. Tech.*, no. 6, 1974, pp. 27–29.
6. G. G. Haselden and L. Klimek, "An Experimental Study of the Use of Mixed Refrigerants for Non-Isothermal Refrigeration," *J. Refrig.*, vol. 1, no. 4, pp. 87–89, May–June 1958.
7. V. Kaiser, C. Becdelievre, and D. Gilbourne, "Mixed Refrigerant for Ethylene," *Hydrocarbon Process.*, vol. 55, no. 10, pp. 129–131, October 1976.
8. D. R. Lascelles and R. S. Jebson, "Some Process Applications of Heat Pumps," *Int. Inst. Refrig. Comm. Meet., Melbourne, 1976.*

# INDEX